The Mound Builders of Ancient North America

Madison School District #38
Educational Services Dept.
5601 North 16th Street
Phoenix, AZ 85016

Additional books by E. Barrie Kavasch:

*The Mound Builders Of Ancient North America:
4000 Years of American Indian Art, Science, Engineering, & Spirituality*, 2004
Ancestral Threads: Weaving Remembrance in Poetry & Essays & Family Folklore, 2003
The Medicine Wheel Garden: Creating Sacred Space for Healing & Celebration, 2002
American Indian Healing Arts: Herbs, Rituals, & Remedies for Every Season of Life, 1999
Enduring Harvests: Native American Foods & Festivals for Every Season, 1995, 2001
Earthwise: American Indian uses of Native Trees, 2000
Hands of Time: Select Poetry & Haiku in Five Seasons, 1999
Native Harvests: American Indian Wild Foods & Recipes, 1977, 1979, 1998
Earthsense: American Indian Ethnobotany, 1996
American Indian Cooking, 1989
Herbal Traditions: Medicinal Plants in American Indian Life, 1982
Botanical Tapestry, 1979

Guide Books:

A Student's Guide to Native American Genealogy, 1996
Guide to Northeastern Wild Edibles, 1981, 1995
Introducing Eastern Wildflowers, 1982, 1995
Guide to Eastern Mushrooms, 1982, 1995

School & Library Books for children & Y/A:

Dream Catcher, 2003
The Seminole: Indian Nations, 2000
Zuni Children & Elders Talk Together, 1999
Crow Children & Elders Talk Together, 1999
Apache Children & Elders Talk Together, 1999
Seminole Children & Elders Talk Together, 1999
Blackfoot Children & Elders Talk Together, 1999
Lakota Sioux Children & Elders Talk Together, 1999
Earthmaker's Lodge: Native American Folklore, Activities, & Foods, 1994

The Mound Builders of Ancient North America

✦

4000 Years of American Indian Art, Science, Engineering, & Spirituality Reflected in Majestic Earthworks & Artifacts

Written & Illustrated by
E. Barrie Kavasch

Author of The Medicine Wheel Garden, American Indian Healing Arts, Enduring Harvests, Native Harvests, & Earthmaker's Lodge

iUniverse, Inc.
New York Lincoln Shanghai

The Mound Builders of Ancient North America
4000 Years of American Indian Art, Science, Engineering, & Spirituality Reflected in Majestic Earthworks & Artifacts

All Rights Reserved © 2004 by E. Barrie Kavasch

No part of this book may be reproduced or transmitted in any form or by any means, graphic, electronic, or mechanical, including photocopying, recording, taping, or by any information storage retrieval system, without the written permission of the publisher.

iUniverse, Inc.

For information address:
iUniverse, Inc.
2021 Pine Lake Road, Suite 100
Lincoln, NE 68512
www.iuniverse.com

ISBN: 0-595-30561-X (pbk)
ISBN: 0-595-66181-5 (cloth)

Printed in the United States of America

A portion of the royalties from this book will benefit the Healing & Ethnobotany Research Projects at the Medicine Wheel Healing Center in Bridgewater, CT.

Ink illustrations © 2004 by E. Barrie Kavasch,
Except where otherwise noted: Jon L. Gibson & the Louisiana Archaeological Survey & Antiquities Commission: 2 site drawings of Poverty Point. Squier and Davis, 1848. Photographs courtesy the Ohio Historical Society; the Illinois Historical Society; Select maps from Robert Silverberg. Artwork courtesy of the National Parks Service

Cover art: © Hopewell Shamans Burial Painting by Louis S. Glanzman
Mound City Group National Monument near Chillicothe, Ohio; artistic reconstruction based upon careful scientific excavation & research.

Kavasch, E. Barrie
   The Mound Builders of Ancient North America/E. Barrie Kavasch
   240 pp cm. Includes biographical references & index
1. Mounds, prehistoric Indian Mounds—History, prehistory, contemporary.
2. Indians of North America—Ancient Mound Builders—cultures, lifeways.
3. Earthworks (Archaeology/Anthropology)—North American cultures.
4. Temple Mounds/Effigy Mounds. 5. American Indians rites & rituals.
6. Sacred Native Practices; engineering, science, burial mounds. 7. Title.

To American Indians everywhere...
Their ancient past and ancestors,
Their many legacies,
Present lives, families, and children,
Their bright futures.

The water bug
　　is drawing the
　　　　shadows of evening
　　　　　　to her on the water.

　　　　　　　　　　　　—a Yaqui song

Contents

Introduction~"Discoveries"..xxi

Chapter 1   The First Americans~The Peopling of North America..1

Chapter 2   Native American Diversity~Cultures Evolve & Spread...10

Chapter 3   Timeline~Native Life in the Americas & Elsewhere in the World...31

Chapter 4   Mound Builders Compared~Adena, Hopewell, Effigy, Mississippian,..................................43

Chapter 5   Adena Mound Building Culture (1000–100 B.C.)..52

Chapter 6   Hopewell Mound Building Culture (200 B.C.–A.D. 500)..62

Chapter 7   Effigy Mound Building Culture (A.D. 350–1300)...79

Chapter 8   The Temple Mound Builders: Mississippian Culture (A.D. 750–1500)....................................93

Chapter 9   A Monumental Task~How Did They Build Those Mounds?...103

Chapter 10   Cahokia~City of the Sun.....................116

Chapter 11   Daily Life~What was Daily Life Like for the Mound Builders?..140

CHAPTER 12 Celebrations and Sacred Rites~Ceremonial &
            Ritual Possibilities 163

CHAPTER 13 Creative Reflections~Legacies from the Ancient
            Mound Builders Worlds..................... 182

CHAPTER 14 Conclusions~How Did These Magnificent
            Cultures End? 195

Glossary... 215
Illustrations 219
For Further Information 223
Mound Sites & Places to visit 227
Author's Profile 239
Index.. 241

Acknowledgments

My greatest respect & admiration is due the ancient mound builders, ancestors of many contemporary American Indians. They continue to expand & enrich my life & spiritual objectivity. The mysterious creators of many thousands of ancient earthworks cast an amazing spell over the land we call America. What a heritage!

My gratitude to the libraries and librarians, state and federal parks sites, and thousands of other mound builders sites, and to their personnel, throughout the country that have helped to make this book possible. I am most grateful to Sheila Tintera, Carl DeMillia, Barbara Nelson, Sue Buckley, Sue Ford, MaryAnn Jackson, Laurie Putnam, at the New Milford Library, and to the great Connecticut InterLibrary Loan Program. Much respect also to my own Burnham Library in Bridgewater for their ongoing help.

My respect to all the scholars in this field, especially Robert Silverberg, Jon Gibson, and professor John Strong. My on going respect to the Institute for American Indian Studies in Washington, CT. Sincere appreciation to my friend Margaret Cooper for enduring ideas, patient editing skills, & being the best "sounding board" I could have. My gratitude also to Maya Cointreau and Will Corey for computer wisdom and help.

Abundant praise and gratitude to my family, especially my sister Kathleen Taylor for enormous help, enthusiasm, and interest, and for providing many well researched details of Wisconsin's Effigy Mound Builders. My gratitude also to my sister and brother-in-law Rebekah and Bill Martin for providing vital background materials and supportive interests on Ohio Mounds Sites. Appreciation to Anita Holmes and Millbrook Press for early visionary interest.

Each person fosters knowledge and strengthens progress toward the goal. I am extremely grateful to each individual who has shared their wisdom with me.

In Remembrance

Wendell Chino, Mescalero Apache Chief & educator
Richard & Terry Chrisjohn, Oneida craftsmen & educators
Claude Medford, Jr., Choctaw-Apache basketmaker & educator
Nanepashamet, Wampanoag educator & artist
Irene Richmond, Snipe Clan Mohawk basketmaker & historian
Sara Ransom, Snipe clan Mohawk basketmaker/beadworker
Tom Ransom, Akwesasne basketmaker & farmer/woodsman
Slow Turtle, Supreme Medicine Man of the Wampanoags
Red Thunder Cloud, Catawba herbalist & educator
Keewaydinoquay, Anishinabe Mide' herbalist & educator

Each one has enriched my life with knowledge through their words & consideration. I am most grateful for the sharings.

Acknowledgements & Appreciations

Meredith Begay (& Family), Mescalero Apache Medicine Woman
Ellyn Bigrope, Mescalero Apache teacher & museum educator
Roy Black Bear (& Family), Oneida silversmith & educator
Greg Borland, Lakota spiritual leader
Marge Bruchac, Abenaki storyteller & educator
Joseph Bruchac (& Family), Abenaki storyteller & writer
Dale Carson, Abenaki artist & educator
Nathanial "Stan" Chee (& Family), Mescalero Apache Medicine Man
Rita & Andrea Chrisjohn (& Family), Oneida artists & educators
Katsi Cook, Mohawk Midwife, herbalist & educator
Linda Coombs, Wampanoag educator & dancer
Barry Dana, Penobscot educator & survivalist
Joel Dancing Fire, Cherokee artist & educator
Big Eagle, Paugussett Chief & beadwork artist
Rita Edaakie (& Family), Zuni museum educator
Eva Geronimo (& Family), Mescalero Apache eldercare worker
Robert Geronimo (& Family), Mescalero Apache elder & rodeo cowboy
Rayna Green, Cherokee folklorist & educator
June Hamilton (& Family), Pawnee educator & dancer
Wendell Deer With Horns, Lakota Pipe Carrier & educator
Myron & Elizabeth Klute, Akwesasne Mohawk healers
Kenneth Little Hawk (& Family), Micmac Mohawk storyteller & musician
Oren Lyons, Onondaga Peace Chief
David Bunn Martine (& Family), Shinnecock artist & museum director
Desiree Mays (& Family), Zuni silversmith & businesswoman
Jo Beth & Ken Mays (& Family), Zuni Counsel Woman & silversmith
Erin Lamb Meeches (& Family), Schaghticoke storyteller
Ramona Peters, Nosapocket, Wampanoag educator & artist
Tom Porter (& Family), Mohawk leader & educator
Tsonakwa, Abenaki storyteller & silversmith
David Richmond (& Family), Snipe Clan Mohawk, educator & historian
Trudie Lamb Richmond (& Family), Schaghticoke historian & storyteller
Lizzie Silversmith (& Family), Seneca herbalist
Josephine Smith (& Family), Shinnecock dancer, educator
Marguerite Smith (& Family), Shinnecock Tribal Counsel
Myra & Geronimo Starr (& Family), Creek Nation

Michael Storm, Seminole Alligator Wrestler & educator
Jake & Judy Swamp, Akwesasne Mohawk Tree of Peace Society
Tall Oak (& Family), Mashantucket Pequot, Narragansett Medicine Man
Gladys Tantaquidgeon (& Family), Mohegan Medicine Woman
Monetta Trepp, Creek Nation businesswoman
Janis Us, Shinnecock Mohawk beadwork artist
Ron & Cherrie Welburn (& Family), Conoy/Cherokee educators
Richard Velky, Schaghticoke Chief
Wunneanatsu, Schaghticoke educator & storyteller

> Some dates and numbers can vary as you read through many different resources on the ancient mound builders. For this reason I use the work of Dr. David Hurst Thomas, curator of North American Archaeology at the American Museum of Natural History in New York City—as the final authority. His book *Exploring Ancient Native America: An Archaeological Guide* (Routledge, 1999) is sited in the back of this book in "Further Reading." He is the foremost contemporary scholar in this field.

Oh the mounds!
Realms of earth & sky,
Visions & sacred geography;
Our ancestors pondered the many
Mysteries, & Spirits continue to tantalize,
Haunting us through amazing engineering feats
Casting long, glancing shadows
Across the lens of memory.

Introduction~ "Discoveries"

○ ○

When I was ten years of age I looked at the land and the rivers, the sky above, and the animals around me, and could not fail to realize that they were made by a great power.

—*Brave Bull, Lakota Sioux medicine man*

Oh, these mounds! Great earthen mounds: some in conical shapes, many huge flat topped platforms, and others linear enclosures. Mounds in the shapes of huge bears, alligators (lizards), eagles (Thunderbirds), men, and snake slither, fly, and stalk across the American landscapes. Were these supernatural figures, religious and spiritual beings, or clan symbols for their creators? Certainly this is sacred geography! The greatest number of mounds is centered along the Mississippi and Ohio Rivers. Various mounds spread across 20 states. More than 10,000 mounds were created in prehistoric Ohio and over 15,000 mounds were constructed in Wisconsin! So many mounds and such mystery surrounding them! Americans have been wondering about the mounds and the mound builders for centuries.

The Spaniard Hernando de Soto (1500-1542) was one of the first Europeans to explore what is now the southeastern United States. During his travels in the 1540s, he encountered tribes related to the Natchez Indians with settlements in the lower Gulf region. The Natchez elevated their ruler, The Great Sun, by carrying him on a platform above the ground and building a great mound for him to live upon. De Soto observed them building "high places" out of earth for their chief's houses. "With the strength of their arms," he said, they piled up "very large quantities of earth and stamp[ed] on it with great force until they…formed a mound from twenty-eight to forty-two feet in height."

What de Soto did not know was that the Natchez were among the last of a long line of mound-building peoples whose once great cultures dominated the American heartlands and bottomlands for 4000 years. It would be a few hundred years before any others would realize this. New waves of Americans were moving into the heartlands, and they had much to discover about its ancestral heritage.

Mysterious Mounds

More than 200 years after the Spanish had first encountered Natchez-related tribes along the southern Mississippi flood plains, European trappers and settlers seeking furs and good farmland began moving west of the Alleghenies into the Ohio River Valley. They discovered beautiful land, broad and lush with rolling prairies, acres upon acres of grasslands, and deep woods filled with game animals and fine trees. Along the way they came across numerous Indian camps and settlements where they stopped to trade and gain information about these new lands. Some of the Shawnee, Illini, and Miami Indians lived in summer camps along the fertile floodplains, where they tended cornfields and fished in the rivers. Other times the new comers encountered bands of hunters and traders, and Indian families in the winter camps making maple syrup and maple sugar.

Amidst these abundant natural resources, the newcomers also discovered strange earthen forms all across the land. There were steep conical mounds and long linear earthworks, like giant snakes, enclosing some mounds. There were huge, steep platform mounds that were flat on top and even mounds in animal and human shapes.

Who had built these mysterious earthworks the newcomers wondered? The Shawnee and Miami Indians living in these regions could not explain who had created them. An old Lenape (Delaware) legend told of how the Lenape and Iroquois People had attacked and driven an ancient powerful nation south of the Ohio River, who left behind the massive mounds. Others believed these were actually Cherokee ancestors, whose legends relate a southeastern migration from the Ohio valley.

These massive earthworks seemed to be very old and otherworldly. Trees, hundreds of years old, grew up through some of them, so many observers surmised that they must have been created by a "lost race" that had vanished before the Shawnee and Miami came to live there. A few even suggested that extraterrestrials had made the great earthworks as landing devices. Most were convinced that the mounds were too sophisticated and the earthworks too mathematically correct to have been made by Indians. Evidently they had not seen or heard about the Natchez' mounds de Soto had come across, or if they had, they did not see the connection between those and the more massive and impressive mounds. Yet all could feel the sacred geography of the mounds!

Reading back over accounts from this period we can see how the myth was born of a lost race of mound builders—a mysterious civilization that had once lived and thrived and then disappeared from the American wilderness. These mounds were unlike anything the early European immigrants had ever seen.

William Cullen Bryant, a romantic poet who surveyed the "mighty mounds," dealt with their origins in this poetic ode to the mound builders in 1832 in *The Prairies:*

> *A race, that long has passed away,*
> *Built them [the mounds];—a disciplined and populous race…*
> *These ample fields*
> *Nourished their harvests, here their herds were fed,*
> *When haply by their stalls the bison lowed,*
> *And bowed his maned shoulder to the yoke.*

Then he continues with an explanation of the mound builders' demise:

> *The red man came—*
> *The roaming hunter tribes, warlike and fierce,*
> *And the mound-builders vanished from the earth.*

Evidently it never occurred to Bryant that *"the roaming hunter tribes"* were actually descendants of the early mound builders. In Robert Silverberg's detailed study The *Mound Builders* (1968), he noted, "Men in search of a myth will usually find one, if they work at it." The early settlers certainly did work at the myths of the mound builders—resisting the common sense notion that thousands of intriguing mounds across America might have been made by earlier Indians who were the ancestors of the Indians they encountered. Centuries ago it was hard to get an objective overview. It would take several hundred years of research to understand the magnitude of all this.

Searching for Clues

Not long after the early trappers and settlers began pushing westward past the Alleghenies, people started to study, map, and excavate many of the mounds in order to find out what might be in them, who had built them, and why. Early scholars and investigators, together with teams of workers, began systematically excavating many mounds well before the birth of archaeology as a science. In some mounds they found human bones and unusual artifacts such as copper ornaments, shell jewelry, and thousands of fine pearls.

Thomas Jefferson, the third United States president, was one of the first people to examine a small conical mound near his home in 1780, while he was governor of Virginia. He found that it was full of human bones that looked as if they had been "emptied...from a bag or basket." He estimated that this mound held perhaps a thousand skeletons and surmised that it was probably a burial mound.

An Ohio physician found almost a thousand skulls in the mounds near the Ohio River in the 1830s, which he studied, trying to learn more about who these people had been. Later a newspaper editor in the 1840s explored more than two hundred mounds and one hundred earthworks in Ohio describing the many skeletons, pottery, priceless ornaments, and other objects his team found.

Yet many puzzling mysteries remained about who the mound builders were, when they lived, and why they built mounds. Apparently the conical mounds were burial places holding hundreds of graves and ornate grave offerings. These were evidently sacred cemeteries wherein dead loved ones were encased in an "earth mountain" pointing up into the Sky World. But what about the platform mounds and the mounds shaped like animals? Investigators called the earthworks "sacred enclosures" honoring Indians' religious purposes. This was especially true of the thousands of earthworks throughout Ohio, Illinois, and Wisconsin

While the earlier investigators raised more questions than they could answer, their work remains valuable because many mounds disappeared during settlement times, before scientific investigations of them could be completed to solve the mysteries of the mounds. Some were excavated and looted by treasure hunters for hundreds of years. Others disappeared beneath the farmers' plows, roadways, and development.

> **Early periods of 'discovery' were driven by people's thirst for knowledge and fed by the compelling mysteries of the mounds. Some individuals worked to save the mounds and their buried remains for history, while others looted them for profit, and many others simply found them in the way of 'progress and development.'** *Most were insensitive to the fact that many mounds were sacred burial sites not to be disturbed.*

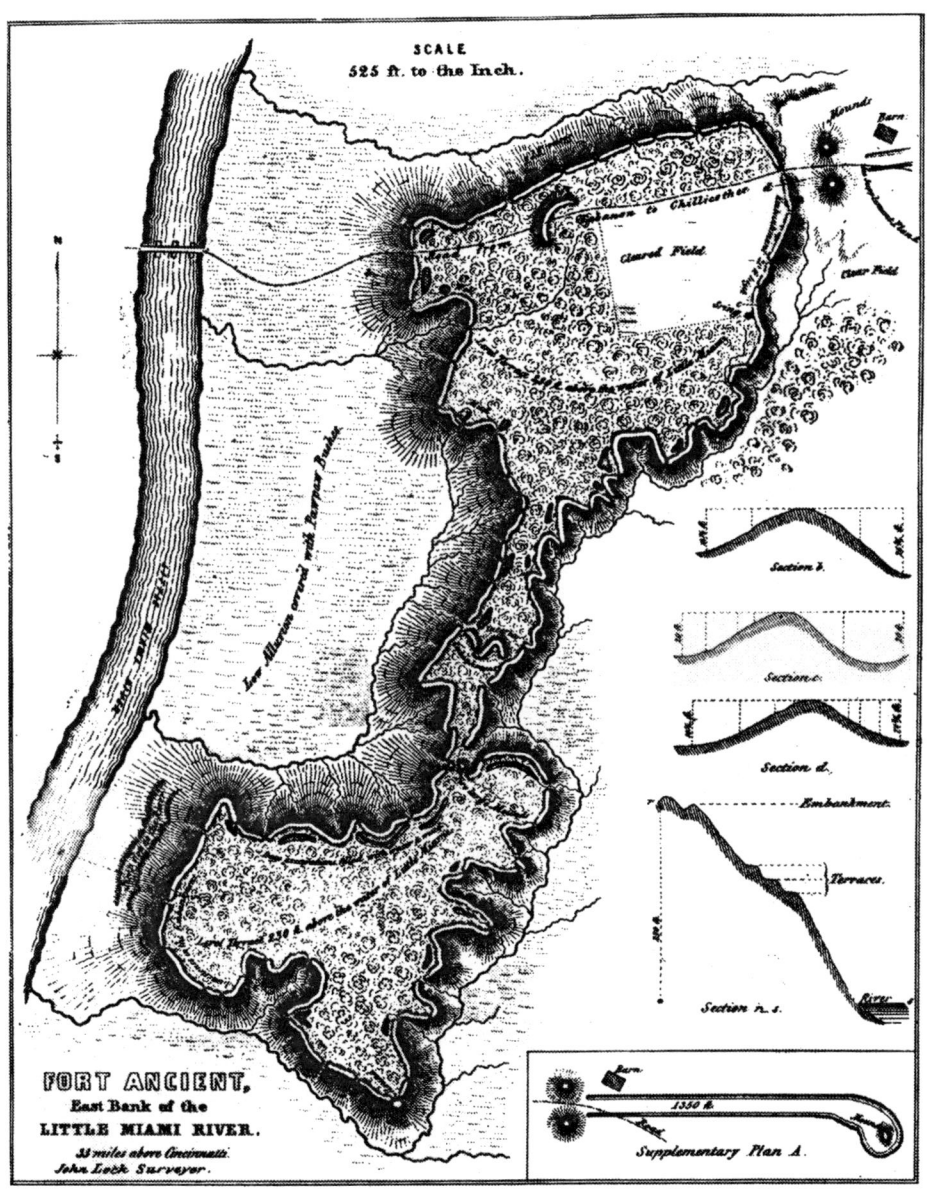

Diagram of Fort Ancient. Squier and Davis. 1848

Scientific Excavation

Aware that the mounds might be totally destroyed before their mysteries were uncovered, in 1881 Congress approved $5000 for a full-scale scientific study of the mounds. Cyrus Thomas, a Smithsonian scientist, and his assistants examined two thousand mounds in twenty-four states and amassed almost 40,000 grave objects, including thousands of skeletons. He published his report in 1894 for the Smithsonian Institution (which gave birth to the *Bureau of American Ethnology Series*), dispelling the theories that the mounds had been built by a mysterious lost race. He was able to show, by detailed examinations of these remains, that the mounds and earthworks had been created by earlier groups of American Indians over a long period of time. The skeletons were definitely those of ancient Indians along with their fabulous sacred grave goods. Careful examination revealed that these mound builders knew how to harvest their many resources and trade across a wide geographic network

Further studies have confirmed Thomas's findings. Archaeologists continue to study the mounds sites trying to learn more about the prehistoric Indians who created them. These specialists have been able to reconstruct some pictures of the mound builders' life and accomplishments. Archaeological studies have been able to date when many mounds were built and how many millions of cubic yards of earth were moved to create them. This science has also revealed that the mound builders belonged to many different prehistoric Indian groups who had striking differences yet shared stunning similarities over time.

But there is much more to know. There still remain many mysteries about the mound builders. Giant earthworks formed diverse temple mounds, altar mounds, burial mounds, effigy mounds, and geometric mound enclosures. These stunning feats of American Indian science, religion, ceremonialism, and engineering continue to radiate awesome energies across the natural landscapes as monuments to dynamic early societies thousands of years after they were initially created.

Wisconsin Effigy Mounds

It appears that in eastern Sauk County, there is a total of 734 mounds, whose existence is fully established...It is definitely known that 337 of these earthworks were constructed as tumuli [burial], one as an enclosure, 183 as effigies and the rest as ridges (with the exception of a few

mounds of unknown shape.) Of the effigies whose shapes are definitely known, there are 43 birds, 47 bears and 12 mink, with the other miscellaneous types. More than 300 of the total number of mounds are now leveled, and are only here recorded by virtue of previous surveys, or other authentic data. A total of 198 mounds still remain undisturbed, and others in various states of distruction (1906).~"A Standard History of Sauk County, Wisconsin, 1(1918:134)

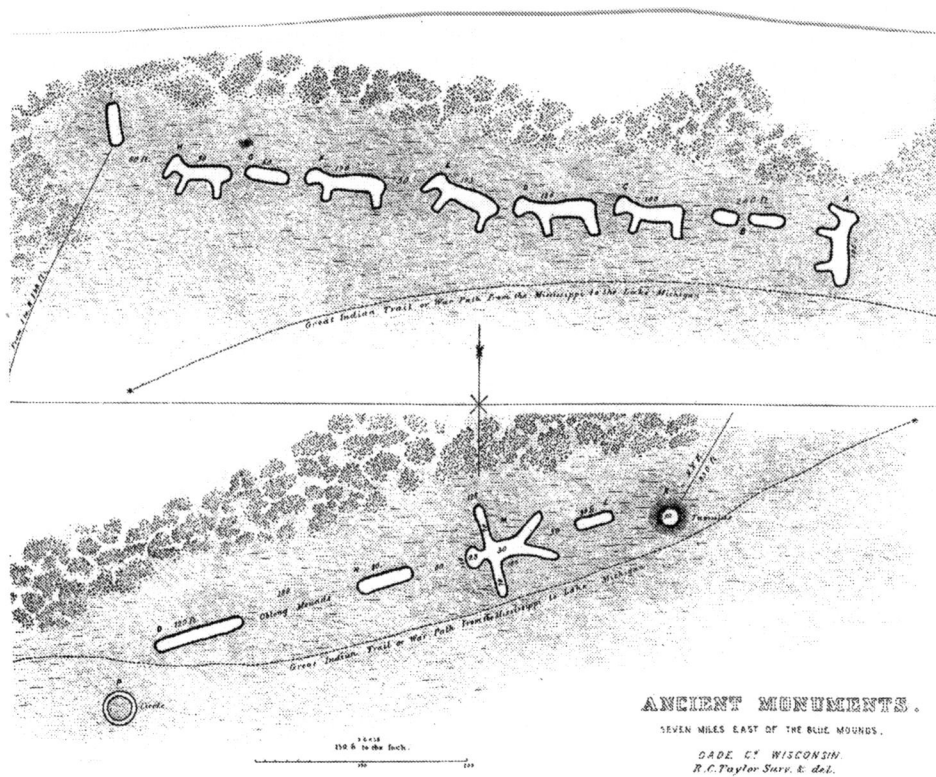

**Ancient Monuments
Diagram of WisconsinEffigyMounds.** Squier and Davis, 1848.

Modern Archaeology

For hundreds of years scholars have excavated prehistoric Indian burial sites in order to better understand these earlier peoples, their diets, health, and lifespans, their differing burial styles, and something about their spirituality and sense of ceremonialism. Finding a burial site was once considered an exciting clue to the past.

Now archaeologists no longer excavate Native American grave sites, except where a site is to be destroyed. It has taken centuries for Americans to come to the consciousness that is so well expressed by Roger and Walter Echo Hawk, Pawnee attorney, historian, and authors of *Battlefields and Burial Grounds*.

Archaeologists and museums have sought out Native American graves in order to study the contents. Archaeologists believe that much can be learned about North America's past from the study of Native dead and the objects buried with them. On the other hand, some Native Americans have strong religious concerns about protecting their dead from disturbance.

It has been more than three hundred years since the Pilgrims opened the first Native American grave. Over the centuries, a double standard regarding the treatment of human graves developed. Disturbance of white dead was regarded with horror, and laws strictly protected white graves. But Indian graves were dug up freely in the name of science.

National legislation was passed in 1990 under the Native American Graves Protection and Repatriation Act (NAGPRA). Both this federal law and state legislation now protect all American Indian burial sites. Now archaeologists get excited about finding ancient campfires and garbage pits, because we can learn more about the people who left them by studying these remains.

Radiocarbon Dating is a unique way of finding the age of ancient objects. This technique (called C-14) was discovered in 1949 by Willard F Libby, a physical chemist, who won the Nobel Prize for his breakthrough. Plants and animals absorb carbon in the form of carbon dioxide throughout their lives. "When an organism dies, no more carbon enters its system, and that which is already present starts its radioactive decay. By measuring the beta emissions from the dead organism, you can compute roughly how long ago that organism died," explains noted archaeologist David Hurst Thomas. Continuing developments have made this form of dating even more accurate over longer periods of time. This is the archaeologists and anthropologists most useful tool for determining the dates of ancient people and their objects.

Prehistoric time refers to that ancient time before history was recorded for those who would follow. The mounds were all built in ancient prehistoric times over a period beginning more than 4000 years before present time.

1

The First Americans~The Peopling of North America

> *The Great Spirit is our Father, but the Earth is our Mother. She nourishes us; that which we put into the ground She returns to us, and healing plants She gives us likewise. If we are wounded we go to our Mother to lay the wounded part against Her to be healed.*
>
> —Big Thunder, Wabanakis Nation, 1900, Maine

> What was North America like for the first humans? What were early prehistoric people like, and how did their cultures change over time? Here we glimpse how the environment and cultures evolved during prehistoric times.

Paleoindian Cultures (13,000–7000 B.C.)

The first Americans were prehistoric hunters, Paleoindians, who walked across the northern "land bridge" that emerged during the last Ice Age. At that time sea levels were much lower than they are today. Paleoindians could travel along windswept ground and packed snow between great ice sheets following large animals. Using long thrusting spears with stone tips, these early peoples moved across North America—from Alaska and the Pacific shores to the Atlantic coasts and down into South America—probably tracking large land animals including the woolly mammoths, mastodons, horses, camels, saber-tooth tigers, dire wolves, peccaries, bears, antelopes, and ground sloths.

The mammoth, a shaggy, longhaired American elephant roamed widely during ancient Ice Age periods as did its equally large cousin, the mastodon. Archaeologists have found remains of Paleoindian sites that date from about 13,000 B.C. Many scholars believe that older sites will one day be found. The Paleoindians who dominated the land until about 6000 B.C. were not one group of people but different groups, many of which developed dynamic cultures based upon their skills, ceremonial practices, treatment of their dead, spiritual traditions, and survival instincts.

> *Paleoindians probably fished and ate a wide variety of seasonal wild foods, from ground squirrels and mice to wild berries, nuts, seeds, and roots. They may also have harvested wild plants and mushrooms for foods and medicines. Traveling in small family units or bands, these early hunters and gatherers probably moved quite regularly following seasonal game animals. They must have camped in caves or beneath rock outcroppings, and they built brush shelters for protection from harsh weather conditions.*

Archaeologists recognize three distinctive cultures within the Paleoindian traditions based on striking differences in their tool making and hunting needs: Clovis culture (9500-8500 B.C.); Folsom culture (9000-8200 B.C.); Late Paleoindian culture (8000-6000 B.C.) These culture groups developed unique stone tools and spearpoints of flint and chert for hunting big game animals.

The Clovis culture was distinguished by the shape and size of the large, flat fluted points (carefully worked stone spearpoints with sharp edges), and other stone tools for butchering game animals. These ancient people found the special stone quarries where they could obtain fine flint (chert and obsidian) to flake into classic spearpoints. Flint knapping was (and is) a precise method for stone-on-stone percussion using hand-held techniques to craft fine tools. These skilled artisans also worked mammoth and mastodon ivory into fine spearpoints, fish spears, and other objects. A dead mammoth or bison supplied tons of meat, bone, and hides for people's needs.

The Folsom culture points were more finely flaked and smaller than the earlier Clovis points. As the climate changed and warmed, the Folsom people mainly hunted bison. Folsom hunters obviously worked together in packs. Archaeologists find that they used the unique land formations on the northern Plains to stampede bison over cliffs in order to kill and butcher enough meat and hides for winter survival needs. Every part of the bison (buffalo) was used: the meat for food, hides for robes and clothing, sinew for stitching hides together and making string, bones for tools and carvings, skulls for ceremonial uses, horns for spoons and rattles, hooves for glue, and dung for fuel.

These Paleoindian groups lived during a period of global warming, between 10,000 and 12,000 years ago, when the large prehistoric animals were becoming extinct, winters became steadily warmer, summers were cooler, and sea levels rose. As the great glaciers melted, coastal areas were flooded, causing shorelines to rise. The huge animals and plants of the Pleistocene era gradually became extinct. Scientists advance several theories suggesting that large animals were over hunted and could not compete and survive the environmental changes. Many smaller animals could adapt more easily to these shifting weather and vegetation patterns, as could the Paleoindian hunters who sought them as prey.

Gradual warming lead to the spread of the modern deciduous forests and marshes by about 7000 B.C. Huge evergreen and deciduous trees began to dominate cooler regions, where springs, rivers, and lakes were abundant. Rolling grasslands and lush prairies beyond the woodlands supported many other animals like buffalo, antelope, elk, and caribou, along with wolves, mountain lions, bobcats, coyotes, beaver, rabbit, prairie dogs, and numerous smaller creatures, which adapted to live in various environments.

The Late Paleoindian or Plano culture was noted for even finer and more diverse stone spearpoints and other types of stone and bone tools. Clearly, Paleoindian people adapted to changing environments as efficient hunters and gatherers, and their populations increased. They probably hunted and snared smaller animals and seasonal fish and gathered wild plants.

The eastern forests stretched from the Great Lakes south to the Gulf of Mexico and east to the Atlantic Coast. Fir, spruce, tamarack, hemlock, and pine trees grew tall and often intermingled with birch, beech, hickory, and oak trees. These lush regions supported dense understory growths of ferns, shrubs, and wildflowers, which in turn supported wild turkey, partridge, quail, and many game birds and animals. Native populations slowly increased and people turned to hunting a range of smaller animals like deer, beaver, mink, raccoon, otter, and fox, in addition to a wide variety of fish.

These environmental changes required prehistoric people to develop different ways of living. As hunting and fishing quarries changed, people modified their weapons. Native People fashioned smaller stone spearpoints and different tools to butcher and skin their game animals. Gradually an Archaic Indian culture developed and replaced the Paleoindian culture. These Native People gathered seasonal foods and prepared and cooked them in new ways. They used large grinding stones to crush seeds, nuts, and dried roots into coarse flour and paste for making breads and soups. They also used hardened clay balls and small rocks, heated in the fire and then dropped into souppots, to heat and cook more foods.

> ### The Peopling of North America
>
> There are several theories about how people first got to America. While most scientists suggest that the first Americans were ancient nomadic hunters who walked across a land bridge from Asia beginning about 25,000 years ago, others say that these Stone Age hunters came in small boats, following the exposed coastlines and drifting on sea currents to put ashore and explore new regions. The Atlantic and Pacific coastlines extended further out into the sea than they do today. Some scholars believe that some of the earliest prehistoric campsites may be submerged beneath the ocean.
>
> Many contemporary American Indian groups have a different explanation altogether. They say that their ancestors have always lived here: *since time beyond memory.* All over America different "Origin Stories" tell how "the First People" emerged from darkness, or water, or fell from the Sky World, or emerged from the Lower World into this one. Their creation stories tell of various ways each tribe came to live on the land we now call America. These are fascinating insights into native wisdom and culture, as you will see in the Chippewa Origins Tale and the Choctaw Creation Story. Native stories generally relate a deep, intrinsic relationship with Earth Mother on many levels.

Archaic Culture (7000–1000 B.C.)

Like their Paleoindian ancestors, the prehistoric people of the archaic period were highly efficient hunters. They were also practical gatherers and fishermen seeking seasonally abundant harvests. With the huge Ice Age animals gone, they hunted large and small game animals and birds. In the deciduous forests of the east, they found deer, bear, moose, wild turkey, raccoon, and fox. In the abundant rivers, streams, and lakes of the eastern woodlands they caught beaver, otters, muskrat, frogs, and crayfish, and they fished for fresh-water fish and mollusks. Along the coasts, they may have caught bluefish, salmon, and an occasional beached whale. They probably gathered lobsters and shellfish, along with a wide variety of nuts, seeds, berries, wild herbs, and mushrooms.

The Archaic people developed new weapons and tools to make their work easier. The ancient atlatl, or spearthrower, attached an extra throwing handle and banner stone weight that helped hunters throw a spear farther and with more force when hunting large game like moose, deer, and bear. Stone toolmakers fashioned and drilled banner stones from basalt cobbles and notched points from flint, chert, or obsidian that they wrapped with wet sinew, made from animal tendons, around the point to hold the spearpoint or tool blade on the handle. They fashioned axes by pecking and grinding rock into shape and hafting it (securing) on a hardwood handle. Archaic people created twined and coiled basketry and used stone mortar and pestle for crushing and grinding seeds and nuts, to make them easier to eat and to digest.

Long spears could be hurled much further and with greater force when attached to a thrusting stick (Atlatl). This provided added efficiency when a weighty banner stone was attached. Here are a few Tennessee banner stones.

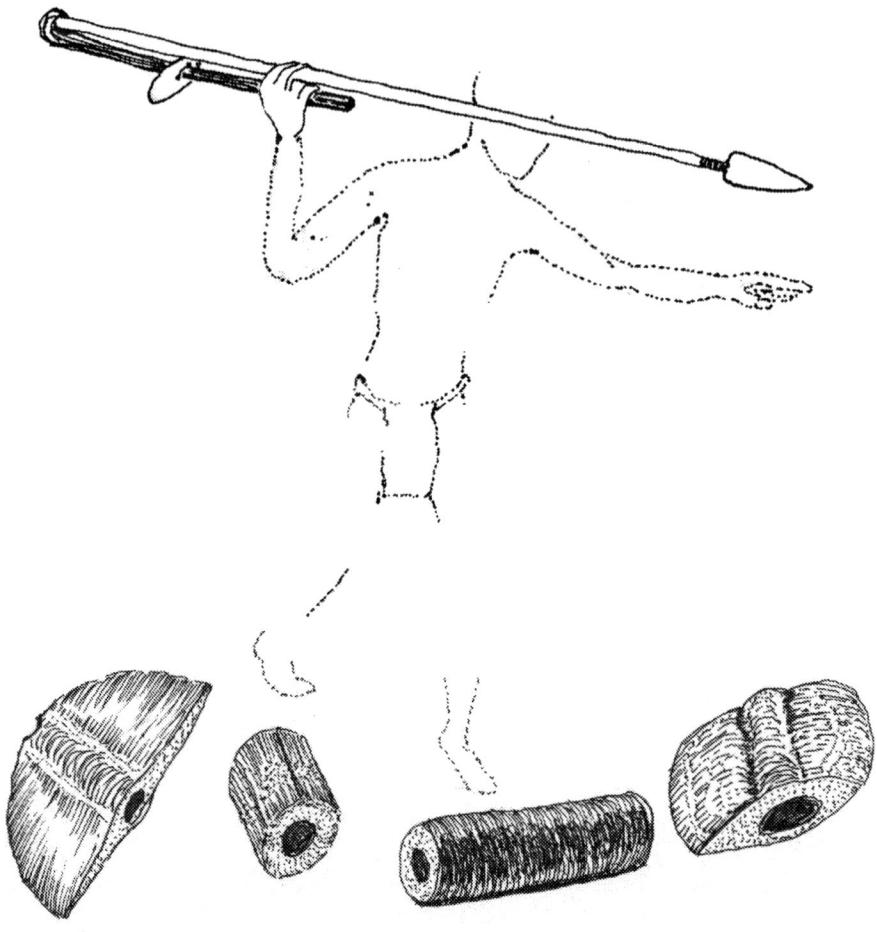

Tennessee banner stones & atlatl–polished drilled stones; University of Tennessee

Archaic Indians lived in small family groups, and in time they began settling in more or less permanent villages, traveling seasonally to hunt, fish, and gather. The Archaic people spread far and wide across this continent and flourished for about three thousand years in North America's heartland, from 7000-1000 B.C. Some Archaic people lived so successfully with their hunter-gatherer traditions that they did not embrace agriculture, especially across the high Plains and the Northwest Coast. The Archaic lifeways lasted almost 10,000 years in these regions, until European contact began. The Archaic lifestyle/culture thrived, but in the eastern woodlands it was gradually replaced by the Woodland Culture about three thousand years ago.

Mississippi River: The Highway of Life

The mighty Mississippi River [*Missis,* great, *Sippi,* river] is as impressive as its Algonquian (Ojibwa) Indian name assures. It flows from headwaters in northern Minnesota's Lake Itasca over 2300 miles to the Gulf of Mexico. If measured from its northwestern tributary in Montana, it flows over 3741 miles to the Gulf of Mexico~nearly the length of the Amazon River in equatorial South America. This meandering river turns back on itself in some places~and drains forty percent of America's heartland. A map of its huge trunkline and many tributaries resembles a great Tree of Life, and it certainly was central to the life of the prehistoric Indians in America's heartlands. This river has influenced the dynamics and prehistory of our continent and continues to exert its dynamic presence as America's longest, largest river. Native People called it "Father of Waters."

The Mississippi drains two Canadian provinces and thirty-one states, carrying about 300 billion gallons of water to the Gulf of Mexico each day. It also dumps 500 million tons of sediment every year at its mouth, causing the Louisiana shoreline to grow by about 300 feet each year. The Mississippi River Valley and drainage area covers almost 1,250,000 square miles~territory nearly equal in size to India, and is a mile wide in some places.

Lake Superior and Lake Michigan are drained by the Illinois River into the northern Mississippi. In the east the Tennessee River drains the southern regions of Virginia and southern Appalachian highlands into the Ohio River, whose tributaries drain western New York and Pennsylvania before entering the Mississippi. In the west the Missouri flows from the northern Rocky Mountains and crosses the Great Plains to enter the Mississippi; below this the Arkansas River links the Colorado with the Mississippi; and below this the Red River flows in from the southern regions of New Mexico, Texas, Arkansas, and Louisiana.

Streams and rivers were America's first highways. Indian people traveled these rivers in dugouts and canoes, and portaged (carried their canoes and goods) over land to avoid rough waterfalls and choppy whitewater rapids. Prehistoric Indians camped and fished along these vast riverine regions. Ancient mound builders created settlements that became cities along the fine network of rivers teeming with life. Rainfall was usually ample in these temperate zones providing abundant wild resources throughout very mixed environments. Rich floodplains were bordered by dense woods, or opened out into rolling prairies as far as one could see. River bottomlands were natural birthplaces for early gardens and became the cradle of prehistoric Indian farming in North America.

The Great Lakes

Five sparkling lakes form a giant geographical stairway, as they seem to step down over 600 feet and reach more than 2,000 miles from the middle of North America to the Atlantic Ocean through Niagara Falls and on to the St. Lawrence River. The water flows from one lake into another in this dynamic freshwater system. French explorers named Lake Superior, *le lac superieur*, and it is the largest and western most of the Great Lakes.

Four of the Great Lakes bear their American Indian names: Erie is Iroquois for *wild cats,* and for the tribe who bore this name, known as *the nation of the cats.* Michigan is Ojibwa for *big water* or *great lake.* Lake Erie narrows at its eastern end spilling into the Niagara River and thundering down the 326-foot vertical drop of Niagara Falls into Lake Ontario. Erie is almost a twin to Lake Huron, which it joins at the Straits of Mackinac. Huron is French for *rough, wild*, which is what the French called the Wyandot Indians, whose own name means *islanders.* Ontario is Iroquois for *beautiful, sparkling water,* and this is the smallest and deepest of the scenic lakes. These surviving tribes may include living remnants of the ancient mound builders.

The Chippewa (Ojibwa), Ottawa, Potawatomi, Menominee, Kickapoo, Winnebago (Ho-Chunk), Sac and Fox were among the Great Lakes tribes. The Sioux Indians recalled their distant origins in the Great Lakes regions, before they migrated out onto the Great Plains. The ancient mound builders were also their ancestors.

This Creation Story from the lands of the Effigy Mound Builders in Wisconsin serves to explain, in part, how the striking geological phenomenon of the Wisconsin Dells, through which the Wisconsin River flows, was formed. Native stories transport us to different realms of understanding.

The Green WaterSpirit

Related from an early telling by Albert Yellow Thunder, (1878-1951) Wisconsin, great grandson of the Winnebago (Ho Chunk) war chief Yellow Thunder.

Back in the earliest times the land was very different from the way we see it today. The People prayed to Earthmaker to give them better hunting grounds. Soon the Creator sent a Green (tco) *WaterSpirit to the lifeless land of snow and ice. The warmth of the green WaterSpirit melted the ice, and with great effort he clawed and bit out channels for streams and lakes across the land to take the collected melt water.*

Then the Green WaterSpirit churned up from his body all the game animals that anyone could ever want. He fired green quills from his skin and they became trees that stood by the thousands from one horizon to the next. This is the way Earthmaker created the Wisconsin Dells, which is called Nic-haki-sutc-ra, *'Where the Cliffs Strike Together'.*

When the Green WaterSpirit had finished his magnificent work, he leapt into the bottomless depths of Devil's Lake (De Wakatcak). *Even though the* Hotcagara *was a great distance away, the enormous sound and shock waves from the Green WaterSpirit's final plunge enabled them to follow the sounds and signs to their new hunting grounds.*

Another Anishinabe version of this relates that the Dells were formed as the Giant Serpent carved the channel of the Wisconsin River as it tried to reach the sea from its home in the great northern forests.

Inspired from Richard L. Dieterle, and from *When the Moon is a Silver Canoe: Legends of the Wisconsin Dells,* by Capt. Don Saunders, 1946. Wisconsin Dells.

Many Native American traditions relate very early migration stories telling how the people came to be where they are. These ancient, mystical stories populate an exciting pantheon of 'teaching tales' and 'creation myths' along with colorful stories of 'how things came to be'. Storytelling is a respected primal teaching pathway keeping traditions and knowledge alive in each new generation of listeners. Selected native stories are included throughout this book along with creative 'imagined' reflections upon prehistoric lifeways, in order to keep the native presence foremost in focus and science in balance.

2

Native American Diversity~Cultures Evolve & Spread

○ ○

Hear me, four quarters of the world~a relative I am! Give me the strength to walk the soft earth, a relative to all that is! Give me the eyes to see and the strength to understand that I may be like you...Great Spirit, Great Spirit, my Grandfather...all over the earth the faces of living things are all alike.

—Black Elk, Lakota Sioux Holy Man

> We seek to understand how early people flourished in the woodlands, plains, and along America's river systems, and created thousands of amazing mounds. Many different peoples used the land and interacted with one another. Meet the early mound builders.

Poverty Point

A spectacular development began during the Late Archaic period in a region that would one day become Poverty Point, Louisiana. Late Archaic Indians there had developed a complex society centered above the fertile flood plain of the Bayou Macon River, a tributary of the lower Mississippi River. (*Bayou* is a Choctaw Indian word for "river," and *macon* comes from the French/Spanish for "large tableland.") Many hundreds of people gradually gathered together to build and live at this site in small thatched houses that simply protected them from the raw environments. They mainly lived and worked outside in this striking center of commerce and life. This early trading center radiated its influences far in all directions.

Around 1800 B.C. these people created a collection of mounds whose outlines formed the shape of a giant bird—a striking effigy with outspread wings standing 70 feet tall at the central mound. It measures 640 by 710 feet. This connects with three smaller mounds and six concentric low ridges arranged like a crescent or partial octagon embracing a curve of the river. Millions of cubic feet of earth were moved to build the six concentric ridges, over two-thirds of a mile long. No one knows the purpose of these mounds or who these people were, yet we marvel at this amazing feat of engineering. Today we call them the Poverty Point Culture. Perhaps their highly evolved religious concepts foreshadowed these designs. As earlier people settled in growing communities, they developed detailed spiritual and ceremonial practices reflecting their beliefs. More than 100 related sites were clustered along the tributaries that fed into the lower Mississippi River, where permanent villages and hunting campsites once flourished during the late Archaic.

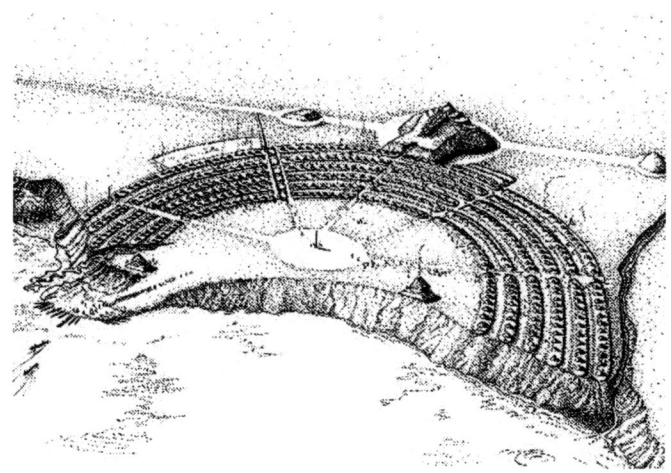

Artists reconstruction of the central ridged enclosure at the Poverty Point site as it may have appeared in 1350 B.C., with a population of perhaps a thousand people. *Drawing by Jon Gibson,* author of *The Ancient Mounds of Poverty Point, 2000.*

> This dynamic prehistoric site covers a 400-acre preserve today. Beyond this elite center, more than a hundred Poverty Point sites have been identified in Louisiana, Arkansas, and Mississippi, and these influences can be traced to early sites in Missouri, Florida, and Tennessee. Countless similar tools and cooking materials were found at these other sites, especially Poverty Point clay figurines, stone beads and pendants, and flints, along with polished stone artifacts. The "ancients" abandoned Poverty Point about 700 B.C. and during the next thousand years new earthen mounds were being built and great networks of trade routes were interlinking them with new cultures.

Objects collected from the Poverty Point site show that these people possessed great craftsmanship and were involved in long-distance trading networks. Here we find objects made of soapstone (steatite) from Georgia and Alabama, lead ore (galena) from Missouri, copper from the Great Lakes, diverse stone for tools from Alabama, Tennessee, Kentucky, Ohio, Indiana, Arkansas, and Mississippi. Poverty Point societies loved exotic materials and crafted them into fine ornaments. They were consummate artists!

Tiny jasper owls, less than an inch high, were drilled with small holes to be worn as beads and pendants. Baked clay balls and small clay figurines recovered from this site might have been made by the thousands and were used for cooking foods, like "artificial stones." This would have allowed these early mound builders to boil water in tightly woven baskets and bark containers. The baked clay balls could be repeatedly baked in a hot fire, then lifted with a bark "shovel" or two long sticks used like tongs to drop the hot balls into containers of water with meats and grains waiting to be heated and cooked. Repeating this fireside process several times over one can make hot soup or stew. This would accompany other foods that were roasted directly over the campfire, and baked in the ashes beside the fire (like simple breads.)

Jasper Owl, one of many found, carved about 1500 B.C.
--Poverty Point

Radiocarbon dates indicate that the earthworks were built between fourteen and eighteen centuries before the birth of Christ. This was an eventful time throughout the world. In Egypt, Amenhotep IV, his queen, Nefertiti, and the boy pharaoh, Tutankhamen, were ruling, and the Canaanites were being enslaved. In Turkey and Syria, the Hittite Empire was expanding. In Iraq, Babylon and its lawgiver king, Hammurabi, were in power. In Crete and surrounding Mediterranean islands, Minoan civilization was reaching its peak. In Britain, Stonehenge was being completed, and in Pakistan, the great planned city of Moenjo-Daro was succumbing to flooding. In China, the Shang dynasty was flourishing, and in Mexico, the Olmec chiefdom was ascending.~Jon L. Gibson, *Poverty Point: A Terminal Archaic Culture of the Lower Mississippi*

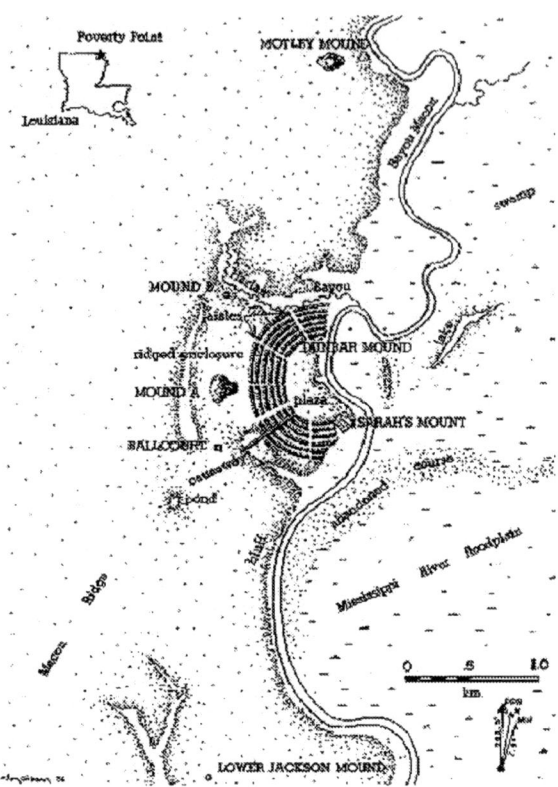

The Poverty Point Earthworks. Drawing depicts the extensive layout, which spans nearly four miles from north to south. *Drawing by Jon Gibson*

Paleoindians and Archaic Indians had been successful hunters and gatherers for thousands of years, adapting to the different geographical and environmental conditions all across North America. As native populations grew, some groups began domesticating select wild plants and settling into villages. Gradually farming spread among many native populations. Imaginative prehistoric people, ancestors of some modern American Indians, constructed hundreds of earthen mounds over the course of two millennia.

Excavations at Russell Cave, Alabama, where prehistoric Indians lived, worked, and overwintered for thousands of years, provide insights into how cultures evolved. Pottery appeared about 4500 years ago in this region and gradually spread throughout the eastern woodlands. Native hunters began making smaller projectile points, suggesting the move from spears to the use of bow and arrow.

Native people developed unique cultural traits that distinguished one group from another over the course of time. These Archaic Cultures were flourishing in prehistoric America at the same time that the Egyptian pyramids were being built along the Nile.

The Woodland Culture (1000 B.C.-700 A.D.)

North America's people and climate continued to change. Thousands of years ago the environments of eastern North America were wilder and more thickly wooded and marshy than they are today. Willows, alders, and dogwood shrubs along with towering canebrakes and cattails filled the lowlands—the sandy bottomlands and marshes. Southern regions supported bottomland hardwood forests lush with redbud, sycamores, oaks, bald cypress, and loblolly pines along the lower Mississippi and its tributaries. In some regions tall and extensive forests developed that covered countless acres. Early naturalists observed in the 1700s that a squirrel could easily travel from tree to tree, from the Atlantic coast to the Mississippi River, and never need to touch the ground. Large herds of deer and great flocks of wild turkeys, geese and ducks populated these fertile environments, which in turn supported growing populations of native people.

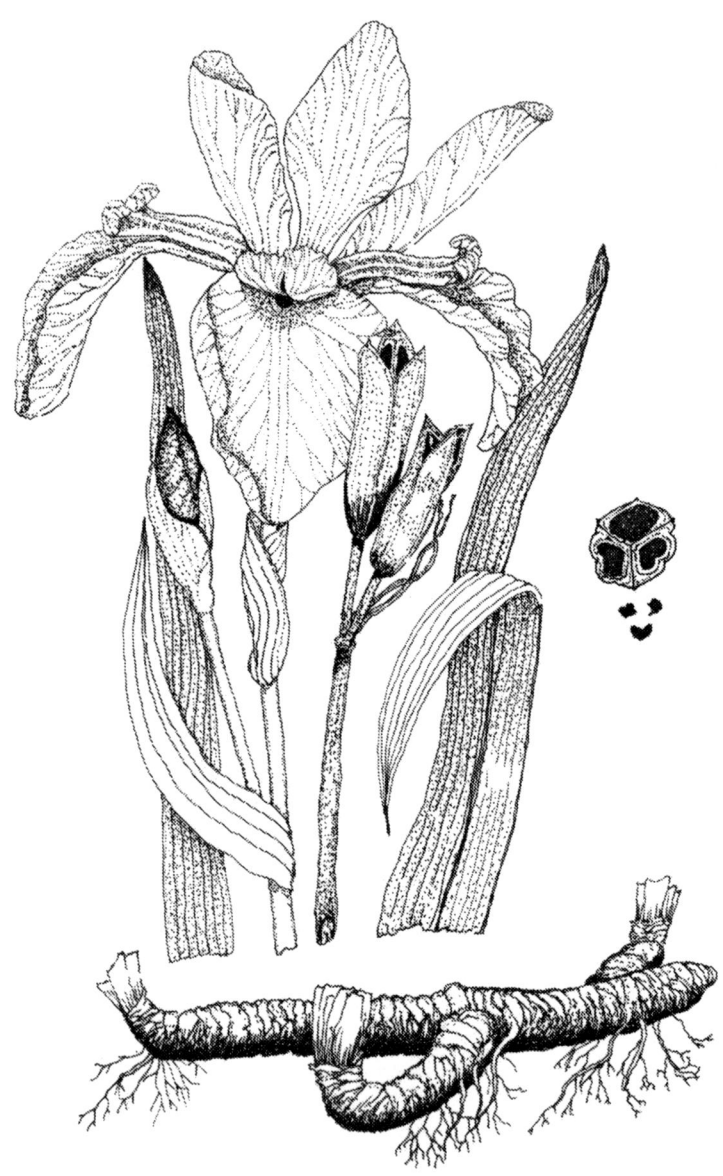

Wild Blue Flag, *Iris versicolor*--valued medicinal herb found commonly growing in open meadows and woodlands, often in great round colonies.

Here too were medicinal colonies of sweet flag, blue flag, bog onions, swamp milkweeds, and wild leeks. The uplands were thick with towering oak, hickory, beech, maple, ash, sassafras, tulip poplar, and walnut trees. Majestic fir, spruce, pine, tamarack, and hemlock trees along with junipers and cedars gave a cool evergreen spirit to the forests. Shady understory growth included a range of edible and medicinal viburnums, hazelnuts, witch hazel, ironwood, arrowwood, and sumac, shading healing herbs such as mayapple, ginseng, ginger, sarsaparilla, hops, wintergreen, wild garlic and onions, partridgeberry, buffaloberry, and a range of wild mushrooms, lichens, and mosses.

Bottomlands along river floodplains and marshes grew thick with alders, willows, dogwoods, and spicebush shrubs along with large econiches of sweet flag, wild iris, and various sedges and rushes. Native orchids and other medicine plants grew throughout these bogs and wet regions. And other varieties of native orchids and medicine plants grew in higher, drier areas, and across windswept grasslands.

The climate in these regions was probably very similar to ours today. Changeable spring weather and long hot summers followed long, severe winters with deep snow. Autumn and then Indian Summer provided dazzling days of cool, mild harvesting weather with lots of changing colors in the deciduous forests. Abundant rainfall, during most seasons, assured lush growth for plants and animals.

Wild American Ginseng, *Panax quinquefolius*–valued medicinal herb of the shaded woodlands, often found growing in colonies; now endangered in the wild.

About 3000 years ago the people of the eastern woodlands began to develop new technologies and to use resources in new ways. Perhaps they had learned more distinctive hunting and craft-making techniques from distant trading partners or chance encounters. These woodland people also developed new ways of living together and honoring the dead. They began to make distinctive pottery, to domesticate plants and become gardeners, and to construct burial mounds. These are some of the hallmarks by which scientists separate the Woodland Indians from their earlier ancestors.

The Woodland Indian cultures brought together many traditions. They were many different groups, bands, and clans of native people. This is an umbrella term that we use to group together people with similar characteristics, as we work to understand them better. Some of their distinctive tools, weapons, artworks and burial practices distinguish one group, or culture, from another.

The woodland people created settled societies in the east. During the late Archaic, people had begun to settle together in more sedentary lifestyles. Early Woodland, Middle Woodland, and Late Woodland Periods are marked by distinctive cultural differences—in the way people made their tools, weapons, pottery, ceremonial works, and their evolving farming and trading capacities. But all shared certain characteristics.

Hunting and Gathering.

Abundant resources were available to the Woodland peoples. Eastern forests provided them with diverse wild animals for food and clothing, and for the most part, these prehistoric people knew where to look for what they needed.

Most years provided rich habitats where Indian people camped and lived during the summer and autumn harvests. Most tribes were also regional opportunists moving with their resources without depleting the things they most depended upon. They moved to hunting and fishing camps, and to coastal and shoreline areas for seasonal fish and shellfish. Enormous shell middens (waste piles where earlier people discarded shells and other food items) tell the story of ancient feasts. They knew how to choose the right fiber and bark for clothing and medicines, wood for fires, and tools, nuts, berries, and herbs for foods. We can better understand how the scientists who study these ancient Indians called them Woodland Cultures. They lived mainly in the woodlands and clearly had a good knowledge of wild edible plants and healing herbs.

Millions of wild fowl migrated along the central corridors from rivers to marshes to lakes, from Canada to the southern Mississippi Valley and Central America providing native people with seasonal abundance in nesting eggs and game birds. Deer, beaver, wild turkey, rabbit, opossum, raccoon, and wild geese and ducks contributed to their seasonal foods. An abundance of fish, shellfish, and frogs also made their diets interesting. Woodlands opened out to grasslands and provided habitats for bison (buffalo), deer, elk, moose, antelope, and countless smaller animals like beaver, fox, lynx, skunk, opossum, mink, marten, weasel, panther, bobcat, mountain lion, coyote, and wolves. Wild turkey, ducks, geese, and swans lived in the marshes and woodlands, along with many other game birds and smaller mammals like rabbits, squirrels, groundhogs, prairie dogs, and mice. The main wild game food resources were probably deer and wild turkey.

Millions of acres of primeval forests were cut by wild "river roads." A major source of food the rivers teamed with large freshwater mussels—which the mound builders collected to eat, and especially for the wealth of pearls found within the mussels. Rivers also supported abundant trout, salmon, white crappie, catfish, bass, and an array of freshwater fish, frogs, crayfish.

Tools and Weapons.

Woodland Indians technologies improved upon their predecessors as the use of the bow and arrow spread. Southern natives fashioned blowguns from cane and sumac stalks for hunting small game and birds. Bone and antler awls and tools increased in sophistication, as did stone tools and weapons. Stone fishnet sinkers and bolas were used more extensively along river and marsh areas.

Transportation and Trade.

Woodland Indians traveled and traded along the system of majestic rivers, fed by many smaller rivers, creeks, and springs which threaded through the lush woodlands, fertile floodplains, broad marshes, bogs, and rolling prairies of ancient America's heartland. The ancient Mississippi river was both "lifeline" and "Mother" for earlier civilizations. She (the Mighty River) could be devastating as well as life giving.

Indians followed well-worn animal trails on hunting, fishing, and trading trips. Yet the rivers were their primary highways. Hollowed-out logs, or dugouts, became the mode of transportation long before the first canoes were made. Huge old trees were "girdled" (their bark removed around the base to kill the cambium layers) and when the trees fell, the interior wood was burned and scraped out until a dugout was created.

The large sturdy dugouts could be used to transport food and other commodities. They also could be packed with considerable trade goods along with several oarsmen and chiefs, who could thread along the river highways from one Indian village to another. Indians established far-flung trade routes to get more exotic materials from other regions, trading with distant tribes in all directions. As diverse Indian groups stockpiled more surplus goods, especially foods and artworks, they had more to trade and barter with other trading partners. Different groups of people have probably always sought ways to trade with each other. They traded stories, foods, customs, and choice articles. The origins of trade and commerce are traced into the distant past. The trade and growth of pottery, the development of agriculture, and the construction of funerary mounds mark Woodland traditions. These three traits are the hallmarks of native life by about 1000 B.C., and earlier in some spots. Populations increased and the farming of plants intensified.

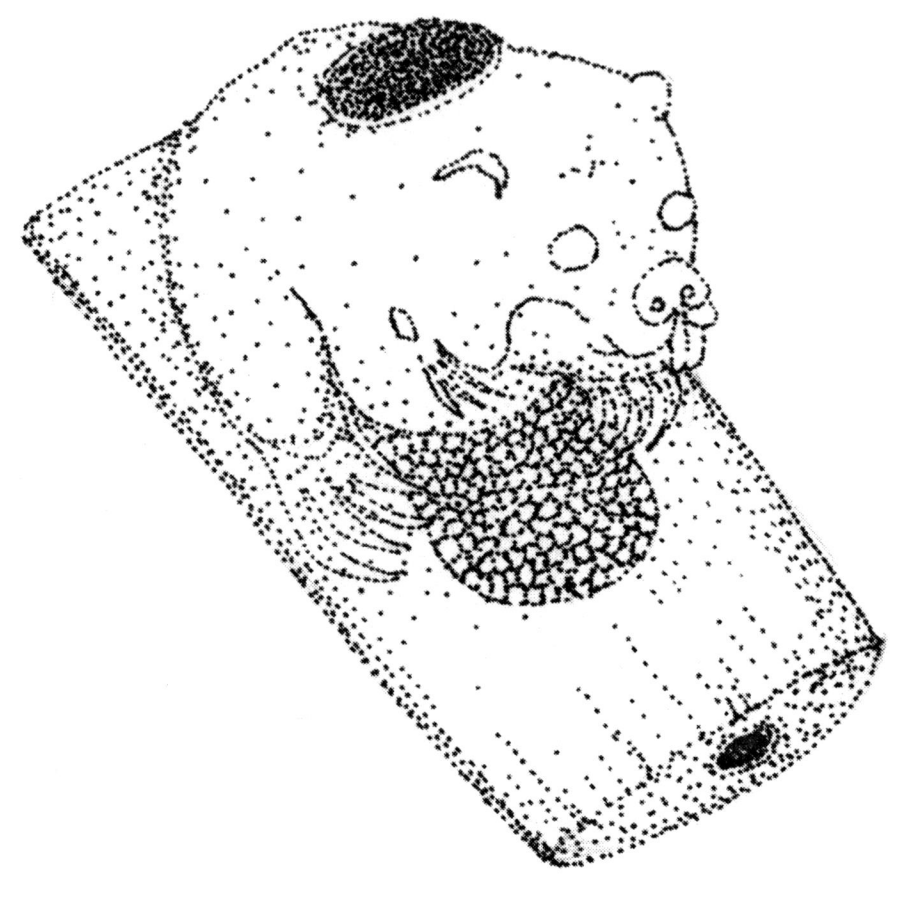

Hopewell Beaver Platform Pipe

Settlements.

Farming demands a more settled life style. As the Indians embraced agriculture, they moved about less and their villages began to grow larger. Woodland economies grew and many prospered. Populations continued to grow and major centers of Indian life emerged during Woodland times because of success with farming and trading. Some Indian communities were very large 3000 years ago. The cultivation of edible plants enabled groups to develop into societies and settle into villages. Villages and farm fields grew into prominent centers in various regions, feeding valuable trade networks and exchanging goods across great distances.

Storage and Agriculture.

As the Woodland Indians became more versatile in their hunting, fishing, and plant collecting in eastern North America by about 1000 B.C., their populations increased. They also began managing the local plants growing wild, like wild gourds, sunflowers, goosefoot, marsh elder (sumpweed), wild leeks, garlic and onions (Chicago is an Indian word for "place of the wild (or stinking) onions), groundnuts, acorns, hickory nuts, and black walnuts. As some groups became more successful than others in gathering and storing certain foods and other items, they began to trade and barter with their neighbors.

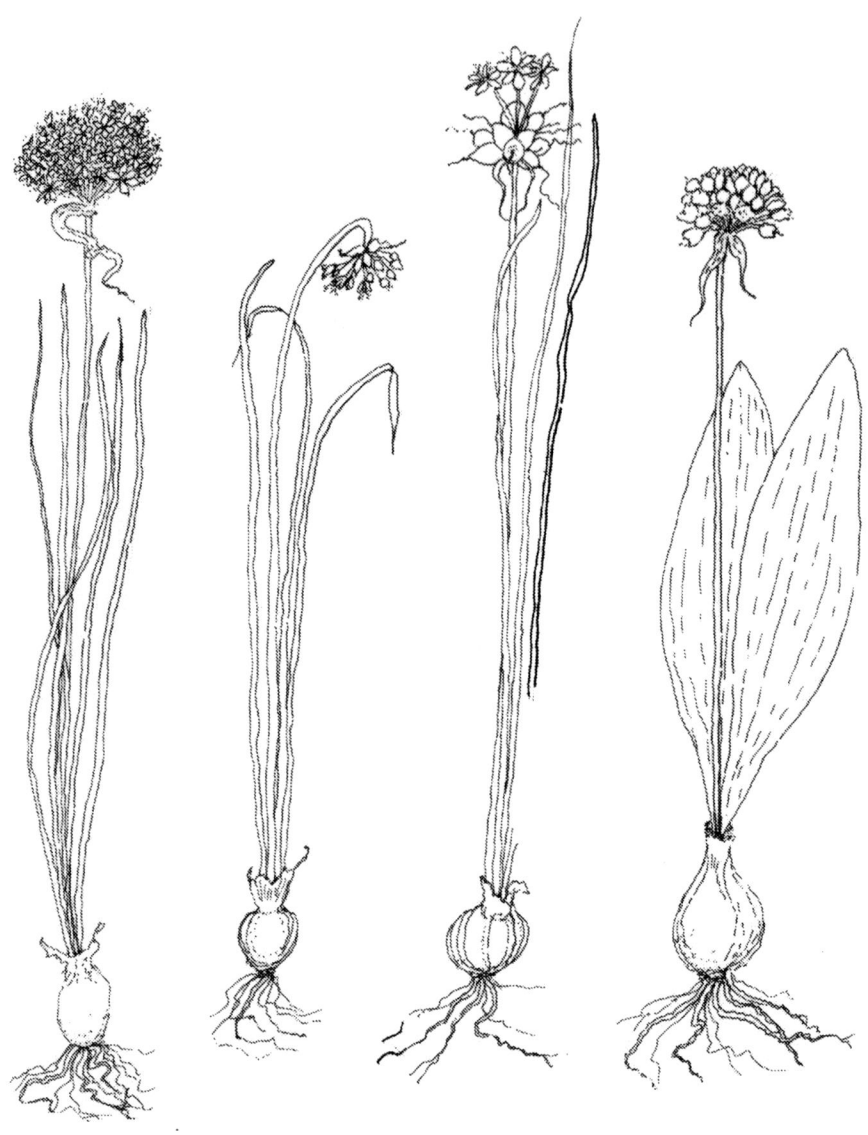

Wild Onions, *Allium stellatum,* **Nodding Onion,** *A. cernuum,* **Garlic,** *A. canadense,* **& Wild Leeks,** *A. tricoccum*–widespread & valuable perennial foods

These talented native people became more involved with the seasonal cycles of annual and perennial plants and mushrooms, saving the most select seeds for replanting. Their floodplain gardens produced rich resources to sustain their growing tribes. While they continued to harvest the woodland bounty and fish the many lakes, rivers, and streams throughout their environments, they also became farmers, who could manage to provide more dependable and ample food supplies. They domesticated various wild plants so successfully that they could live in more permanent settlements in the fertile floodplains along the major river systems, the main highways of life.

As their own food collecting became more efficient, the Indians also learned to store food for the future in large underground pits and protect themselves from periods of lean, "starving times." Huge storage pits were dug and lined with charcoal and rushes to prevent mold, mildew, and burrowing predators. Foods like wild rice, corn, beans, and sunflower seeds were carefully dried and packed into deerskin bags, bark containers, gourd and pottery vessels. These valuable resources were stacked into underground pantries, covered with large sheets of bark and layers of soil. The earth served both as an oven, to cook in, and a refrigerator, to use for cold storage, as well as the ultimate resting place for the dead.

The mound builders must have enjoyed a varied seasonal diet of game and fish, supplemented with delicious wild edible plants, mushrooms, and eventually garden corn, squash, and beans—the "sacred triad" or "three sisters." The three main types of garden vegetables were considered "female spirits and female vegetables" that liked to grow in companionship with one another. Women tended and "owned" the fields and were assisted by the children, who found pleasure in working in Mother Earth.

Religion and Art.

The prehistoric woodland people had a compassionate spirituality regarding death and the Afterlife, which affected the ways they buried their dead. The construction of funeral mounds around and over the bodies of their dead was an amazing feat of engineering and human energy thousands of years ago. Many early burials included costly and beautiful grave goods, which were doubtless gifts to accompany the departing spirits into the afterlife. These concepts evolved and changed over hundreds of years. Particular changes enable us to know different cultures more intimately and create the dividing lines that separate one culture from another.

As some Woodland groups became more affluent from farming and trading, their art and ceremonial works reflected greater fineness. Many created amazing mounds to commemorate this, like the many conical burial mounds that they constructed pointing ever higher into the Sky World.

The Mound Builders

Mound building began during the Archaic Period with Poverty Point and Russell Cave, and continued with the early Woodland Period, intensifying during the Middle and Late Woodland and Mississippian Periods with various technical changes. Two distinguished mound-building cultures with distinctive artwork and burial practices developed during the Woodland Period. The **Adena culture** began in the Early Woodland Period, about 1000 B.C. and lasted until A.D. 100. They were centered in the region we know as Ohio and West Virginia, where they created more than 500 sites. The Adena culture created a new American ceremonialism that would sweep across the country. Burial practices suggest that it was an egalitarian society in which all people were treated fairly equally. The Grave Creek Mound, in Moundsville, WV, which rose over 70-feet high with a flattened top 60 feet in diameter, is considered one of the largest prehistoric Indian burial mounds of its kind in the world.

> *The Adena folk were unusually tall and powerfully built; women over six feet tall and men approaching heights of seven feet have been discovered. It would seem that a band of strikingly different people of great presence and majesty had forced their way into the Ohio Valley from somewhere about 1000 B.C.*
>
> —Robert Silverberg, 1968

The **Hopewell culture** developed in the Middle Woodland Period just overlapping the Adena, from about 200 B.C. to A.D. 500. These people were centered in the Ohio Valley, although their many sites covered eastern and Midwestern North America, from the Great Lakes regions in the north to the Gulf of Mexico in the south. They certainly must have influenced other Woodland groups with whom they traded for exotic materials all across the continent. Hopewellians built many more mounds and carefully engineered more mound complexes than the Adena. The name comes from archaeological excavations in the 1890s at the farm of Mordecai Hopewell, west of Chillicothe, Ohio. Sometimes a major name comes out of a person's efforts or the place where a particular interest began. (A group of mounds is still protected there in a 13-acre square enclosure.)

Hopewellians included many different groups of Woodland Indians whose settlements stretched from the regions of Virginia to Nebraska, and from Mississippi to Minnesota. This was a broad network of diverse people who represented different cultures and spoke different languages, yet they were drawn together by a set of beliefs and symbols that unified them. Some of their striking symbols of a bird talon and a human hand, created from sheet mica, and a buffalo fish (which may have symbolized the underworld), and a hawk (which must have symbolized the upper world and clever hunting abilities) cut from native copper still haunt us today.

Hopewellian Indians in Florida

South Florida was another early Hopewell mound building center with manmade canals that connected the early Ortona Indians with their trading partners almost 2000 years ago. Indians traveled mainly by dugout canoe through this lush subtropical environment and crisscrossed the Everglades hunting and trading. A village covering two square miles at the heart of this network was well engineered around a series of 20 mounds and large ponds. Centered around Lake Okeechobee's mounds, this system reached out to the Caloosahatchee River and the Gulf of Mexico and well beyond.

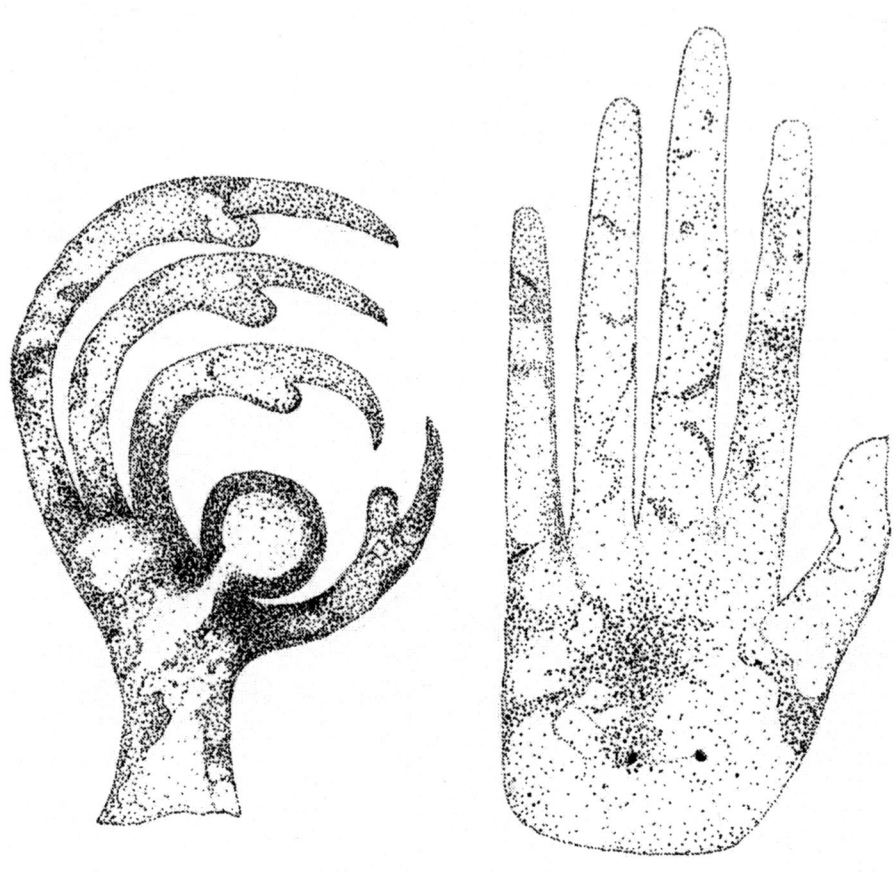

Hopewell Powerful Bird Talon of Mica & Hopewell Mica Hand
Both are powerful shamanic symbols. Ohio Historical Society, Columbus, Ohio

Effigy Mounds (A.D. 350–1300)

In the late Woodland Period, around A.D. 700, mound building shifted. Enigmatic effigy mounds began to be created in some regions of the upper Mississippi Valley. Construction of effigy mounds from about 350 until A.D. 1300 may have reflected changing attitudes toward ancestors. Perhaps clan symbols and guardians were personified in effigy mounds. Were they built to protect ancestors' spirits in the afterlife? Perhaps this presaged a new worldview that evolved beyond the Hopewell beliefs.

Archaeologists called this group the Effigy Mound Culture, for lack of a better term. These northern people continued to construct conical burial mounds and linear mounds, and they also constructed huge mounds in the shapes of bison, birds, panther, lizards, eagles, bears, turtles, and people. There were no artifacts or skeletal remains, for the most part, in the effigy mounds. Along with the Great Serpent Mound in Ohio, hundreds of these mounds crossed western Wisconsin, southwestern Minnesota, and northeastern Iowa. Perhaps 10,000 effigy mounds were once constructed throughout this special region, but many have been obliterated by farming and settlement expansion.

Effigy mounds may have served as gigantic territorial markers or enhanced sacred ceremonial ground where people met for major events. Surely the animal effigies might represent clan totems, unifying related family groups. Some authorities suggest that the effigy mounds echoed the celestial events of sun, moon, planets, and stars vital to the native culture, who devised the crisp effigy mounds as calendars for festivals and planting times, intertwining their cultural life with that of the Sky World. The symbolic architecture of the ten Marching Bears (at Effigy Mounds National Monument in Marquette, Iowa) precisely shadows the Big Dipper in early spring as it marches around Polaris, the North Star. Native traditional healers suggest that the effigy mounds enhanced the power for healing and communication with the ancestor spirits buried nearby. Certainly these are places of considerable power. The full significance of the effigy mounds remains a mystery.

Eventually the Effigy Mound builders evolved into a new society or were displaced about A.D. 1400 by intensive farming tribes who lived in larger villages. Their native copper and obsidian tools, and the many haunting effigy mounds, continue to remind us of the mysterious societies that flourished for over 900 years.

The Mississippians (A.D. 750–1500)

A new group of mound builders emerged around A.D. 750 and took mound building to new heights. Scientists call these mound builders the Mississippians because their culture emerged in the Tennessee, Cumberland, and Mississippi River Valleys.

The Mississippians flourished for about 800 years. Their culture far over shadowed any that came before it. The Mississippians were efficient farmers and master mound builders. They also built great cities and established far-reaching trading networks. Their influence spread throughout the eastern woodlands and beyond. The Mississippians are often referred to as the Temple Mound Builders because of their practice of building great flat-topped mounds upon which they placed temples for their leaders and for periodic ceremonies designed to honor the earth and nature and the people's survival needs.

Mississippian societies distinguished between people of different abilities in a hierarchy governed by economic, social, and religious status as well as by occupation. A whole level of social organization evolved in contrast to earlier Native American societies in these regions. In order to build great earthen mounds, local chiefs had to be able to organize labor. The number obviously varied greatly between small outlying villages with one or two mounds and grand mound centers like Cahokia. Perhaps this variation was faintly similar to our modern states.

3

Timeline~Native Life in the Americas & Elsewhere in the World

∘ ∘

The rocks are Manitou Asseinah, literally "spirit rocks" where God abides.

—*old Eastern Algonquian saying*

> A Timeline overview of the Mound and Temple Mound Builders, and myriad Native cultures before and after them, places them in relation to one another. Our earliest ancestors hunted across ancient times. The past 5,000 years in Indian America to the present tells much about changing cultures, which were quite different from one geographical region to the next. The dynamic Mound Builders began to flourish more than 4000 years ago, dominating Middle America from about 2500 B.C. until A.D. 1600

The technologies and engineering feats of these many cultures continue to astonish us. Our modern world is still dazzled by the amazing achievements. Ancient Indian architects across the Western Hemisphere left abundant signs and symbols of their beliefs and artistic visions in extensive earthworks created without the use of the wheel. The imagination, spirituality, and political dimensions that drove the Mound Builders to create massive earthworks, temples, and villages haunt us.

Many American Indian names continue to live all across our country, like *Missouri*, a tribal name meaning, "muddy waters." More than half the states and their rivers bear original American Indian names, or some form of them. How ancient are the origins of these names? Have these native names come from Mound Builders' origins? Earlier people lived close to the earth and held a deeper perspective of the natural world. Their practical and spiritual lives radiated intrinsic earth knowledge on both a minute and a gargantuan scale that continue to surprise and mystify our sophisticated civilization.

Timeline: Ancient Mound Builders and Native American History

50,000 B.C.	Paleo-Siberian people began crossing the Bering Strait land bridge.
13,000 B.C.	Earliest evidence of Paleoindian life in North America.
10,000–8000 B.C.	Last Great Ice Age (the Pleistocene Period) ended.
9500–8500 B.C.	Clovis Indian culture period
9000–8200 B.C.	Folsom Indian culture period
8000–6000 B.C.	Late Paleoindian (or Plano) culture period
7000 B.C.+	Cultivation of various species of wild plants began
4000 B.C.	Archaic culture period evolved from the Paleoindian culture.
2500 B.C.	Russell Cave site showed earliest pottery; bow and arrow.
1800–500 B.C.	Poverty Point Mounds Complex flourished, then declined.
1000 B.C.	Woodland Indian culture traditions emerged from the Archaic culture.
1400 B.C.	Growth of Eastern Mound Builders cultures:
1500–400 B.C.	Olmec civilization flourished in Mesoamerica, then faded.
1000–100 B.C.	Adena culture flourished across the eastern American midland.
300 B.C.	Early Mayan civilization developed in Mesoamerica
200 B.C.–A.D. 500	Hopewell culture flourished across eastern America, then faded.
200 B.C.–A.D. 600	Nazca civilization flourished in southern Peru, then faded.
A.D. 300	Classical Mayan civilization began to develop in Mesoamerica
A.D. 350–1300	Effigy Mounds Culture flourished, then faded.

Timeline–Native Life in the Americas & Elsewhere in the World

A.D. 750–1600s	Mississippian culture flourished, then faded.
A.D. 700–1400	City of Cahokia housed nearly 20,000 inhabitants by 1100.
A.D. 900–1100	Chaco Canyon towns were built in the Desert Southwest.
A.D. 900–1100	Ocmulgee Temple Mounds flourished in Georgia.
A.D. 900–1200	Monks Mound was built in several stages at Cahokia.
A.D. 900–1200	Toltec empire rose to power, then declined, in Mexico.
A.D. 900–1600	Angel Mounds complex flourished, then declined in Illinois.
A.D. 1000	Woodhenges, circular sun calendars, were constructed at Cahokia.
A.D. 1000–1300	Knife River Indian villages and mounds flourished in North Dakota.
A.D. 1000–1450	Spiro Mounds complex flourished in Oklahoma.
A.D. 1000–1500	Moundville Ceremonial Center in Alabama flourished.
A.D. 1100	Cliff Palace was built in Mesa Verde
1050–1150	Height of the Mississippian culture.
1000	Exploration of Northeastern coastline by Norse and Viking People.
1150–1350	The Southern Cult (Southeastern Ceremonial Complex) flourished.
1200s	Sun Watch Mississippian community flourished on the Great Miami River in Ohio
1345	Aztec city of Tenochtitlan, "place of the prickly pear cactus," was founded.
1492	Eastern seaboard of North America was further explored by Europeans.
1500	European diseases began ravaging North American Indian populations.
1502	Aztec empire reached its greatest extent in Mexico.

1519	Spanish explorer Cortes invaded Mexico with an army.
1540s	The Southern Cult (again) flourishes across the southeastern regions.
1541	Hernando de Soto, Spanish explorer, 'discovered' the Mississippi River, then died.
1607	English settlers established Jamestown in the colony of Virginia.
1620	Pilgrim settlers established Plimoth Plantation in the colony of Massachusetts.
1682	La Salle's voyage of exploration up the Mississippi River for the French.
1729	The Natchez Indians revolted against the French and were eventually decimated.

While Elsewhere in the World...Global Parallels in Ancient Cultures

We are deeply involved in the spirits of all forms of life, not just human life.
—Slow Turtle (John Peters), Wampanoag Supreme Medicine Man

> In unique places around the world, ancient prehistoric people were creating sacred objects and sacred space on a mammoth scale. What challenged these early civilizations to cope with seemingly overwhelming tasks? What were their spiritual linchpins? What remains is tantalizing.

People have always been inspired to build monuments to honor their heroes and gods, and especially their dead loved ones and respected leaders. A culture's worldview, or cosmology, is personified in some of its most prominent constructions. Unfortunately, we have only remnants of some of these earlier societies and their monuments to greatness. Other world cultures also buried their dead in earthen mounds.

Reflections on Distant Polynesian Island Cultures

At the same time that Cahokia and the Temple Mound Builders were flourishing in North America, the *Rapa Nui* people of Easter Island established themselves and their own unique culture on this isolated Pacific Island. The artistic Polynesians settled their distant island about A.D.600, and from 700 to 1600 they carved about 1,000 giant stone figures called *moai* that stood over 30 feet tall. They used their most abundant natural resource, volcanic stone quarried from *Rano Raraku,* an extinct volcano that dominates the island.

These great stone "chiefs" were somehow moved to the coast where they were stood upright, facing inland, on great stone altars. They were believed to channel the energy of the gods back into the island. The *Rapa Nui* also honored a birdman creator called *Makemake.* They carved this likeness in stone and wood all over the island. There are striking parallels here to the birdmen dancers (eagle, hawk, and thunderbird) of the ancient mound builders, and the birdman god of the ancient Egyptians, which are so mysterious and well pictured in their native artworks. It would seem that there was an ancient worldwide movement to create big memorials and honor great birds of prey.

The Great Wall of China

While not a spiritual but rather a protective construction, the Great Wall was first a legendary idea beginning in the 3rd century B.C. The need for fortifications against Mongol raiders caused various northern provinces to begin creating a great wall that would run along the northern border of the country for 1,500 miles (2,400km). It was built of pounded layers of earth alternating with stones and wood. Earthen walls were created from 403 until 221 B.C. by various warring states before China was unified. It was an incomplete network in many areas and never completely joined together.

During the Ming Dynasty a new Great Wall was begun in the 15th century A.D. around Beijing and leading out into the country. Around the capital the walls measure up to 25 feet high and up to 30 feet wide, tapering from the base to the top. The Great Wall was built to impress, as well as to fortify some regions, but it was never completed. Still it is one of the great legacies associated with China. And here is another great culture devoted to massive earthworks during the times when Native American mound builders were constructing great earthworks and fortifications.

Ancient Egyptians

The ancient Egyptians built more than 90 royal pyramids from about 2630 B.C. until about 1530 B.C. All pyramids aligned with the cardinal directions; their sides ran almost exactly due north-south and east-west. The pyramids were all built on desert plateaus on the west bank of the Nile, behind which the sun set. Egyptians believed that a dead monarch's spirit left the body and traveled through the sky with the sun each day. When the sun set royal spirits settled back into their pyramids to renew themselves.

Most of the early pyramids were built of cut stone that was carefully moved and set in place through precise human feats of planning, engineering, and teamwork. Later pyramids were built of mud bricks, also carefully engineered and executed by group labor. Scientists believe that the pyramids, tombs for their kings and queens, were central to the people's ongoing religious activities. Egyptian grave goods and hieroglyphs (picture writing) were essential to safeguard the dead monarch's soul during its passage into the afterlife. These reflected the important hymns, instructions, magic spells, and guidance on how to act in front of the gods.

> It is fascinating to reflect on these similarities on the other side of the world from the ancient Mound Builders. Egyptian pyramid building was declining at just about the time that the Mound Builders were beginning to flourish in ancient America. What spiritual beliefs caused remarkably different cultures to construct such astonishing tombs to hold the remains of their elite? Did they communicate through dreams and visions?

Ancient Olmec, Maya, Toltec, and Aztec Civilizations

The ancient Olmec, Maya, Toltec, and Aztec civilizations of Mexico and Mesoamerica built great pyramids and temples of stone surrounded by immense open plazas and earthen mounds. These became major trading centers while each empire flourished. Scientists see strong parallels and influences between these Mesoamerican cultures and the Mound Builders. Possibly these were also vital trading partners.

The mysterious Olmec developed a civilization in the Mexican heartlands. These amazing artists and ceremonialists carved colossal heads in native stone and constructed hundreds of earthen mounds around a ceremonial courtyard. Their

Great Pyramid at La Venta was built over 100 feet high and more than four times this dimension across its base. The Olmec developed far-reaching trade networks that fed this ancient empire from about 1200 B.C. until around 400 B.C.

The Mayan civilization ruled from about 300 B.C. until A.D. 1541, after the Spanish conquest of Guatemala and Yucatan. The Mayan people created towering temple-pyramids in great stone cities, which were independent states from one ruler to the next. A warrior king, whose lineage passed down from father to son, ruled each city-state. These rulers claimed descent from the gods and inscribed their ancestry on walls, stelae (huge standing stone monuments), and pyramids in celebrations of their powers. Mayan people continue to flourish today and many live in North America now.

The Toltec empire stretched from A.D. 900 until about 1200. These mysterious people were fierce and warlike, ruling much of central Mexico in the 11th and 12th centuries. They built colossal stone figures and temples atop pyramids. The Aztec considered them their ancestors. Scholars believe they influenced the Southern Cult.

"The Toltecs were wise. Their words were all good, all perfect, all wonderful, all marvelous; their houses beautiful, tiled in mosaics, stuccoed, smoothed, very marvelous...They were thinkers, for they originated the year count, the day count; they established the way in which the night, the day, would work." The Spanish priest, Bernardino de Sahagun, wrote these words reported by an Aztec prisoner in 1521.

The Aztec empire flourished from A.D. 1345 until 1521, when the Spanish soldier Cortes and his armies decimated the Aztecs. One of his soldiers, Bernal Diaz, commented when first seeing Tenochtitlan on Lake Texcoco, "When we saw so many cities and villages built both on the water and on dry land...we could not resist our admiration...because of the high towers, cues [pyramids] and other buildings, all of masonry, which rose from the water. Some of our soldiers asked if [it] was not a dream. There was a great market near the ceremonial center of Tenochtitlan, "place of the prickly pear cactus," where 60,000 people came daily to trade countless exotic and handmade wares.

The Mound Builders of ancient North America must have traded, intermarried, and been greatly influenced by some of these early people. Perhaps they influenced each other in more ways than we can know, especially through dreams, visions, and shamanic visits.

The Nazca Lines in Peru

Just as the Hopewell Culture was emerging in eastern North America, the Nazca Culture began developing far to the south more than 2000 years ago. The Nazca civilization flourished in southern coastal Peru between 200 BC and AD 600. They spread over more than 10,000 square miles centered in the Nazca and Ica River valleys. These people developed from the earlier Paracas society, whose religious and cultural traditions they retained. Their economies developed around intensive farming with irrigation in these desert regions. Nazca engineers dug deep wells called *pukios* with interconnecting tunnels to various underground aquifers. The villages and independent chiefdoms shared common religious practices, which focused on rituals designed to insure abundant crops. They also fished the rugged Pacific coastal regions.

Nazca religious art portrays a wide range of fantastic half-human, half-animal creatures, thought to be symbols of the most fearsome creatures inhabiting the earth, sky, and water. The largest ceremonial center was *Cahuachi,* uninhabited it covered more than 370 acres, had more than 40 stepped pyramid mounds that supported ceremonial temples and was surrounded by extensive burials. Shamanic rituals were vital to Nazca life.

Nazca artists created stunning polychrome pottery with naturalistic designs that often incorporated up to 15 different colors. Designs of animals, plants, and fish often accompanied by complex mythological motifs adorned double-spouted bottles, whistling pots, bowls, and vases. Nazcans also wove fine textiles, exquisitely embroidered, and fashioned fine featherwork, and gold and copper metalwork.

Yet perhaps these ancient South Americans are best known for their distinctive lines: huge earth drawings called geoglyphs depicting natural images of spiders, a dog, a hummingbird, killer whales, and a gigantic bird with a 900-foot wind span, as well as various geometric forms. Lines were created by sweeping away surface rocks to expose lighter sandstone underneath. The famous Nazca Lines were once considered mysterious, otherworldly, and thought to serve astronomical functions. It is now believed that the straight lines served as ritual pathways connecting sacred places, and the striking Nazca drawings were apparently created to honor the sky gods who would view them.

Some amazing parallels existed between the Nazca and the Mound Builders, although they inhabited strikingly different environments many thousands of miles apart. Each constructed burial mounds and ceremonial centers, often with straight pathways leading into them and connecting various centers with one another. Each culture developed exquisite artworks that pictured both the natural world and elements of the spiritual and ceremonial worlds. They were farmers. Also each culture possessed the engineering talents and technical skills to construct enormous animal and geometric figures on their landscapes to honor the ancestors and communicate with the Sky World.

Ancient Medicine Wheels in North America

Great stone circles are feats of ritual architecture found across North America, and some are more than 10,000 years old. They were given the generic name of medicine wheels by Western explorers more than a century ago, because they resemble the cross within a circle design created and worn by many Plains Indians. Hundreds of different medicine wheels, large and small, were created, from native stones moved to form huge circles, spirals, and spokes or crosses enclosed within circles. Some of these with central stone cairns (mounds) are the earliest burial sites we know of in North America. Authorities believe that certain medicine wheel sites were also used as solar calendars to indicate seasonal planting, hunting, and ceremonial times. They certainly provide commanding views of the eastern sunrise, direction of sacred origins.

These ancient ruins have a certain spell about them that affects everyone who journeys to see them. Most medicine wheel sites are located in high, remote places. The most famous is the Big Horn Medicine Wheel located high in the Big Horn Mountains (about 10,500-feet) near Sheridan, Wyoming. Ancestors of the Crow, Arapaho, Sioux, Shoshone, and Cheyenne Indians built it about 2500 years ago as a huge earth altar. The 28 rows of stone spokes within the circle count out the lunar cycle.

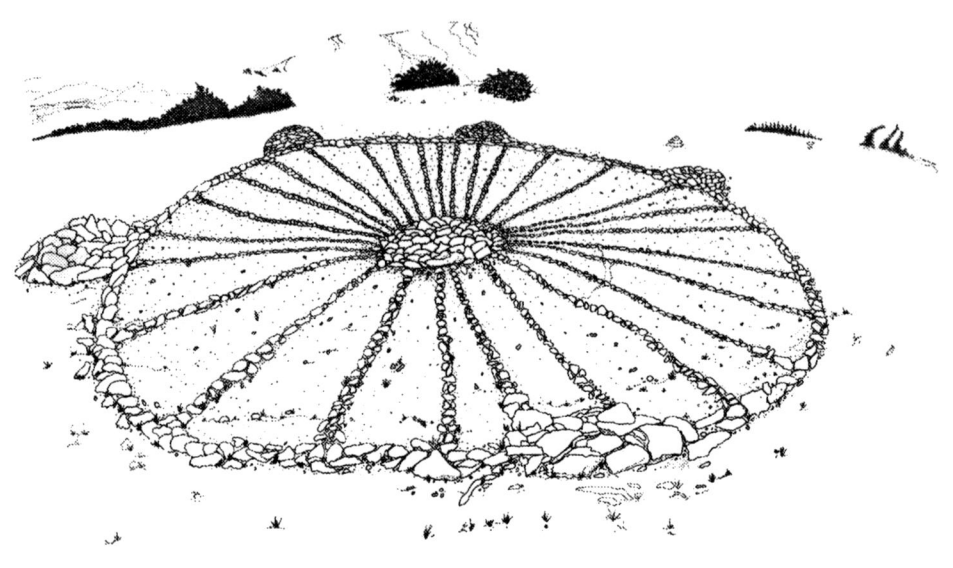

The Big Horn Medicine Wheel in the Big Horn Mountains of Wyoming

Circles have always been sacred basic forms in American Indian dwellings, dances, clothing, and artworks, as well as in rituals and healing practices. Drums, dream catchers, rattles, and bullroarers embody circles mirroring the shape of the sun, moon, and earth. The cycles of life and death are viewed also as continuous circles.

Sacred Symbols

The cross inside the circle design must have had numerous sacred meanings. It is repeated in shell, stone, wood, and pottery objects of mound builders' artwork. The sacred cross surely symbolized the four cardinal directions of north, south, east, and west. It also signifies the four logs of the sacred fire and the ends of the physical world tethered at distant horizons to the upper and lower worlds. The circle embracing the cross also represents the radiant sun, and the moon. This motif in its entirety certainly symbolizes balance.

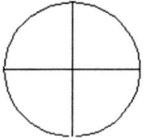

Think not of yourselves, O chiefs, or of your own generation, but think of your children, your grandchildren and those yet unborn whose faces are still beneath the earth.

—The Peacemaker, founder of the Iroquois Confederacy

THE FIRST FIRE (A WOODLAND INDIAN STORY)

The First Fire is an old Woodland Indian story about the origin of fire and how the animals worked hard to bring the first glowing coals to this part of the world. Difficult, almost impossible, work! No one knew how to handle fire, or that you couldn't. It is told in the imaginative way an old storyteller might use to interest young listeners gathered around an evening campfire eager to *know where fire came from*. Each tribal group might have a different story about how the first fire was obtained. Mound builders probably told a similar version of this story around their evening campfires.

"Fire will keep us warm and light up the night," said spider in the earliest days so long ago. *"If we bring it here and take good care of it, fire will be our friend. Shall I go get it?" "Whoo—whoo—who, you?"* mocked owl. *"Your legs are long and crooked. It would take you a bear's sleep to go and return. I shall go myself."* And, owl flew off to the other end of the great land—to the Island of Fire, but singed her feet and feathers as she tried to pick up a glowing coal, and returned without fire. Her feathers continue to show the burned marks to this day.

Then rattlesnake boasted, *"I have tough skin, I shall go fetch fire."* But he was driven back by the flames, which burned beautiful designs in his skin. Each animal in turn offered and went to the Island of Fire attempting to bring back a glowing coal to make the first fire. One by one they returned hurt and tired—without fire.

At last spider departed, spinning her long silken threads as she went…She worked quickly when she reached the Island of Fire, wrapping fire's brightest ember in a silk web, while dancing her magic spider dance and praying to the Creator for success. She hurried back home spinning from one silk thread to another along her silken pathway and carrying the glowing ember in her web sack well beneath her.

All of the animals gathered around her as she placed the glowing ember on a nest of twigs and fire shot up merrily. "Thank you, spider!" They rejoiced in its warmth and the way it could light the night. They took turns feeding it, and they never forgot their gratitude to spider. Each blazing sunset and rosy predawn sky reminds us of spider's patient work and quiet success.

> Grandmother Spider is an honored Creation figure. A Cherokee story tells how spider carried embers to them for the first fire across a web from the Fire God. A Shoshone story tells how spider taught them the art of weaving, as does the Navajo, who say that Grandmother Spider is still weaving the threads of life together.

Southern tribes suggest it was the black Dirt-Dauber Wasp who succeeded in carrying an ember back in her 'clay pot.' Some say that spider also formed a clay pot to hold the first ember of fire. These stories coordinate the earliest fire with the birth of pottery.

4

Mound Builders Compared~Adena, Hopewell, Effigy, Mississippian,...

○ ○

The country knows. If you do wrong things to it, the whole country knows. It feels what is happening to it. I guess everything is connected together somehow...

—*A Koyukon Indian saying, from central Alaska.*

> The Adena, Hopewell, Effigy, and Mississippian cultures had striking differences based upon their technologies and cultural characteristics. It is important to understand how Mound Builders flourished and created thousands of amazing mounds. How did these people live and worship and celebrate life and death? And where was each culture centered? Maps show the great range of these early cultures.

Ancient Mound Building Cultures

> The Adena, Hopewell, Effigy, and Mississippians—the Mound Builders—developed an advanced technology that enabled them to surpass earlier groups. They created new ways of increasing food production. They acquired pottery-making techniques and better ways of storing food. They established trade networks more complex and far reaching than those of their predecessors. They developed engineering skills that enabled them to build enormous earthen mounds in many shapes and sizes.

The Mound Builders formed larger settlement areas where people could express themselves and their creativities in more profound ways regarding the ceremonial world. Evidently some interacted with people of Mesoamerica and were influenced by their ceremonial practices, farming, pottery making, and mound building complexes. In addition, these groups had acquired a compassionate spirituality regarding death and the Afterlife. The construction and ceremonial use of burial mounds led to Native American societies gathering into larger communities. They developed new social systems and ways of governing. These concepts evolved and changed over hundreds of years as they evolved out of the woodland phase. Particular changes enable us to know different cultures more intimately and create dividing lines that separate one culture from another.

Archaeologists' close observations of mound construction, along with agriculture and pottery, show the defining characteristics that distinguished woodland cultures from one another and from the cultures that followed them. As we look back across many hundreds of years of prehistoric time, we use certain keys such as these to help us know who these ancient people really were. Each major culture was represented by many different native groups: tribes, bands, clans, and unique individuals who may have had different languages and lifeways yet shared certain key concepts in common that helped to define them, thousands of years later, as part of a particular culture.

The Natural World in Balance

A sexual division of labor marked a common thread through most native communities. Women's work centered on the village and the surrounding fields in a balanced and complementary series of roles. Women were the growers, creators, nurturers, as they raised crops, bore children, tended hearth fires, and created clothes. Men's work, for the most part, took them away from the villages for hunting, fishing, trading, and fighting. Men and women obviously felt their own special kinship with other beings in the native cosmos. Women created and men killed. They shared equal power in their natural world. Each worked to help preserve the community's oral traditions and keep life in balance.

Childbirth usually took place away from the men and family in order to protect them from the powerful forces present during this sacred time. Likewise menstruating women usually remained apart from the community for several days each month during their time of purification. Many clans maintained birthing huts and menstrual huts for these highly specialized needs when women were at their highest power.

> **Spirits of Nature**
>
> Native Americans have long believed that everything possesses a spirit and deserves respect. These traditions brought them closer to nature in many holistic ways. Living close to the land, native mound builders could eat, sleep, and breathe it. Mother Earth was the creative life force for everything. The plants and animals that populated the mound builders' world were an intimate part of the web of life. Honoring the nature spirits assured a successful coexistence. Honoring the spirits of food and medicines enhanced life and assured greater health. Many Native Americans continue to honor and believe in these traditions today. We call this belief *animism*, a religion practiced in many parts of the world.

The Southern Cult

Between A.D. 1150 and 1350 a Southeastern Ceremonial Complex spread across much of the East with distinctive symbols of its own designs. This was a huge network of people concentrated in three different centers: Moundville in Alabama, Etowah and Ocmulgee in Georgia, and Spiro in Oklahoma. There seems to have been a high degree of social and artistic interaction between these major centers. Evidence shows that the Southern Cult flourished again in the 1450s through the 1600s in response to the Spanish invasions.

Southern Cult symbolism radiates out through sun circles, the cross, the hand and forked eye motifs, conch shell gorgets and cups finely etched, and effigy pipes that immortalize important beings, especially the shamans. These people obviously revered the serpent, falcon, woodpecker, raccoon, beaver, deer, panther, and other wildlife that populated their environments and filled their myths, stories, and religious practices with deeper earth wise meanings.

Moundville Archaeological Park covers 320 acres on the south bank of the Black Warrior River in Moundville, Alabama. This early metropolis features 20 oval and square platform mounds from 3 to 58 feet tall, arranged around a square ceremonial plaza. Death was an inescapable period of transformation here for both the deceased and those left behind, which this society celebrated with amazing intensity.

Populated by skilled artists and trading elite, this society was supported by cornfields and a bustling ritual center, and fed by an extensive trading network. A large palisade surrounding the massive earthworks crowned by temples and council houses protected this site. The elite also lived atop the great platform mounds and ruled more than 3,000 people within Moundville. They controlled smaller centers along nearby stream and river systems. These were probably the early ancestors of the Creek Indians.

Spiro Mounds in eastern Oklahoma were built on fertile uplands along the Arkansas River near the Ozark highlands and the Quachita Mountains, where four huge mounds were created in a north/south alignment. The largest mound stood 33 feet high and 112 feet in diameter; the adjacent mounds were about one half this height and about 188 feet long.

This impressive religious and political center was named after a nearby present-day community. It flourished from 850 to 1450 and its mounds and villages covered about 140 acres. Spiro was obviously a magnificent trading center during its heyday, where many Mississippians gathered. This was the hub of **the Southern Cult**, a center for southern trade links between ancient Mexico and the western Gulf regions.

Unfortunately this great site was significantly damaged by relic hunters, vandalism, and farming. These ancient Mississippians were probably the early ancestors of the Caddo and Wichita tribes.

Other minor sites of the Southern Cult reveal more details about the range of these prehistoric societies. A huge platform mound at **Nanih Waiya Historical Site** in Mississippi is considered the Great Mother in Choctaw creation stories and the ancestral home of the tribe. This mound was constructed about A.D. 400. The Choctaw still hold their annual Green Corn Dances here each year. They call this place "Mother."

Mound Builders Compared~Adena, Hopewell, Effigy, Mississippian,... 47

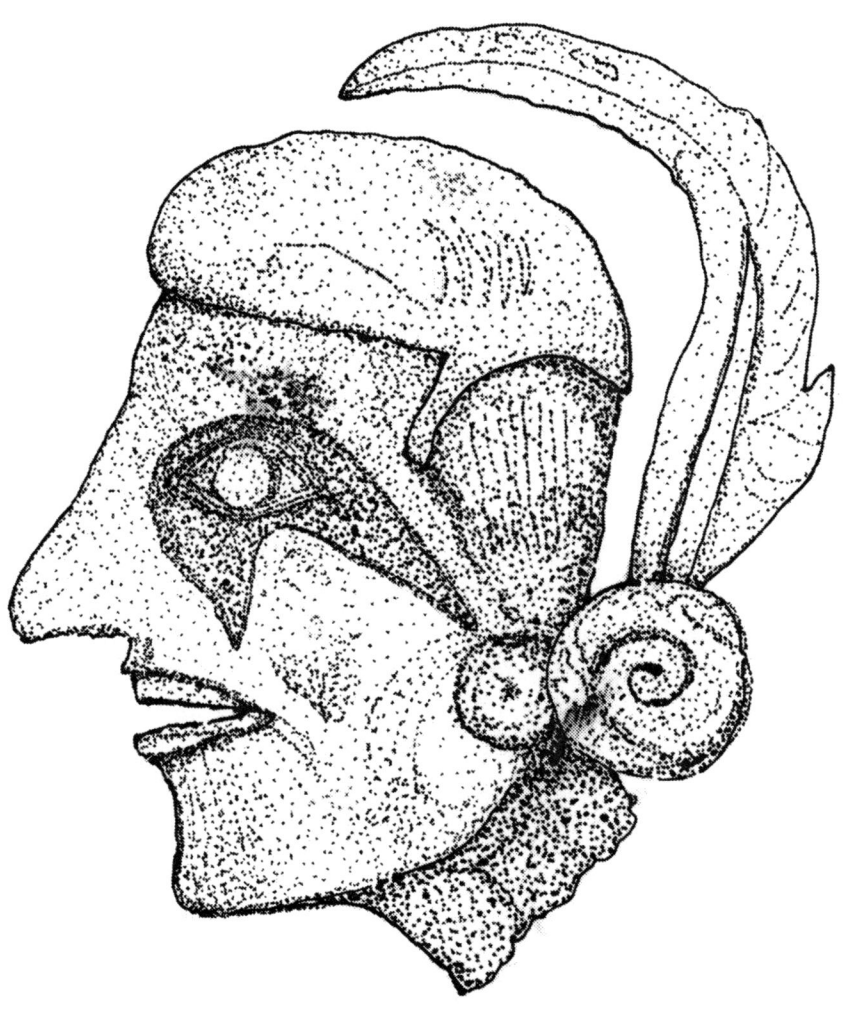

Spiro Mounds Southern Cult Repousse' Warrior

The Natchez Indians

The last great Mississippian culture of temple mound builders lived in settled villages along the lower Mississippi River Valley. These were the Natchez Indians. The Natchez dominated the numerous native palisaded towns and villages along the lower Mississippi and as far west as the Texas border. Many were encountered by the de Soto expedition between 1539 and 1542.

> *"The Governor [de Soto] marched two days through the country of Casqui, before coming to the town where the cacique [chief] was...*[This was noted on June 19, 1541.] *The Governor entered Pacaha, and took quarters in the town where the cacique was accustomed to reside. It was enclosed and very large. In the towers and palisades were many loopholes. There was much dried maize, and the new maize was in great quantity throughout the fields. At the distance of half a league to a league off were large towns, all of them surrounded by stockades..."*

The French settled among the Natchez in the late 1600s and eventually destroyed them, yet they recorded many aspects of Natchez lifeways. The Natchez had a central temple mound near an open plaza surrounded by satellite mounds—the nobility lived upon the central platform (temple) mound tops, and other mounds were the burial monuments to dead leaders, nobility, and their loved ones.

The Natchez had a complex caste system beneath the royal family regulated by behavior and relationships. The Great Sun, the supreme ruler, lived on one of the most central temple mounds along with his family. His mother, White Woman, lived on another temple mound and was one of his close advisors. The other nearby temple mounds were inhabited by his brothers, the Suns, and sisters, Woman Suns, from whom the priests and war chiefs were chosen. The nobles and lesser nobles lived in village homes beneath the mounds, along with the commoners, whom they called the "stinkards." All men and women of the nobility were permitted to marry only commoners. Yet when a noble person died, his or her mates and other close associates were expected to give up their lives and follow the leader into the Afterlife. They followed the ancient Southern Cult religion and were feared as mystics by the French priests and settlers, who wanted the fertile lands for their own plantations.

The Natchez Revolt

Following La Salle's voyage of exploration in 1682, the French began settling in Natchez territories. (Sieur de La Salle (1643-1687), was known as Robert Cavalier.) As the French increasingly coveted Indian settlement locations and rich farmlands, acts of violence occurred between them. The Great Sun's peaceful and well-loved brother, the Tattooed Serpent, served as peacemaker for years, but he died in 1725. The French, who were with them at the time, noted the sacrifices made when this honored war chief died. Many people sacrificed themselves and died with him. They included two of his wives, his one sister, his healer, his first warrior, his head servant and speaker, whose wife also accompanied him in death. His nurse and a man who made war clubs also sacrificed themselves. This poignant account relays their sophisticated spiritual inclinations. The French tried to persuade one of his wives not to sacrifice herself, but she said:

He [Tattooed-Serpent] is in the country of the spirits, and in two days I will go to join him and will tell him that I have seen your hearts shake at the sight of his dead body. Do not grieve. We will be friends for a much longer time in the country of the spirits than in this, because one does not die there again. It is always fine weather; one is never hungry, because nothing is wanting to live better in this country. Men do not make war there anymore, because they make only one nation. I am going and leave my children without any father or mother. When you see them, Frenchmen, remember that you have loved their father and mother and that you ought not to repulse the children of the one who has always been the true friend of the French.

But following Tattooed Serpent's death the violence increased. The Natchez Indians rebelled against the French living in their midst in 1729, when Sieur Chepart, the Louisiana governor, ordered the Natchez Great Village to be evacuated so that he could build his plantation there.

Natchez bands struck and killed almost 250 French settlers and captured their Fort Rosalie in late autumn. The French retaliated, sending two invasions out of New Orleans to attack and kill the Natchez. Many were captured and sold into slavery; some survivors settled among neighboring Cherokee, Creek, Choctaw, and Chickasaw tribes. Other small bands of Natchez hid out along the Mississippi and continued their resistance against the French settlers.

The French and Indian Wars (1754–63) were fought between the English and French and their various Indian allies. This further decimated American Indian populations and set one tribe against another.

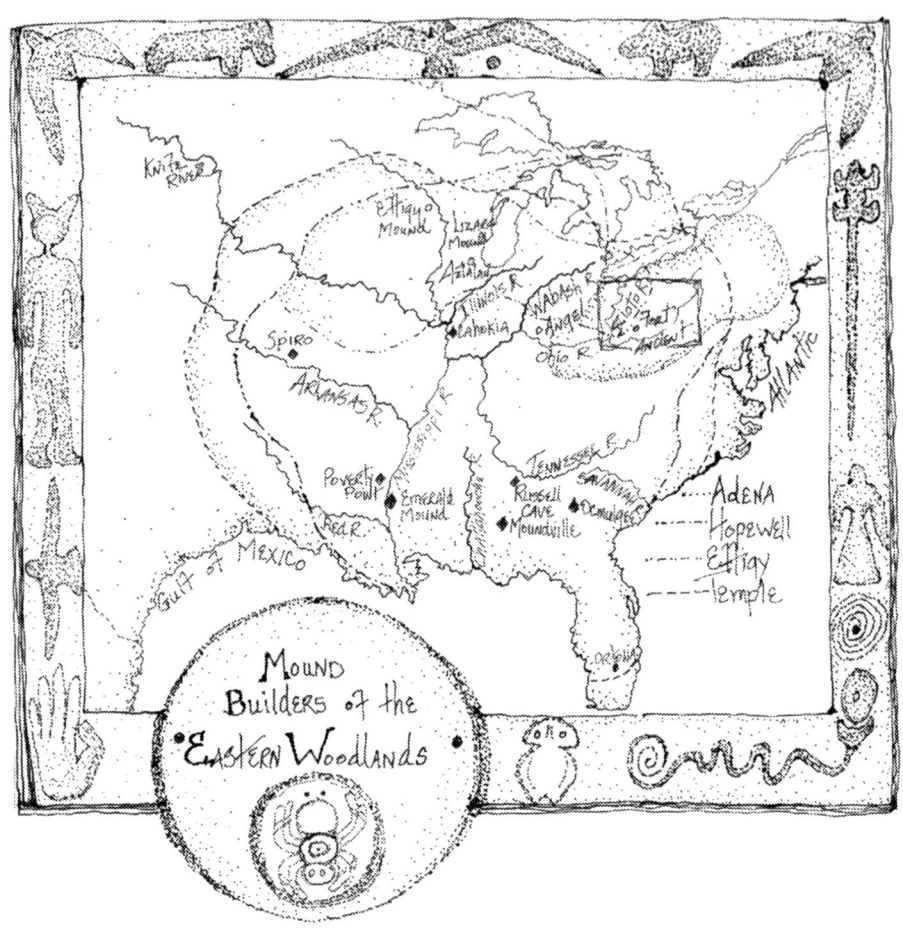

Mound Builders of the Eastern Woodlands Map

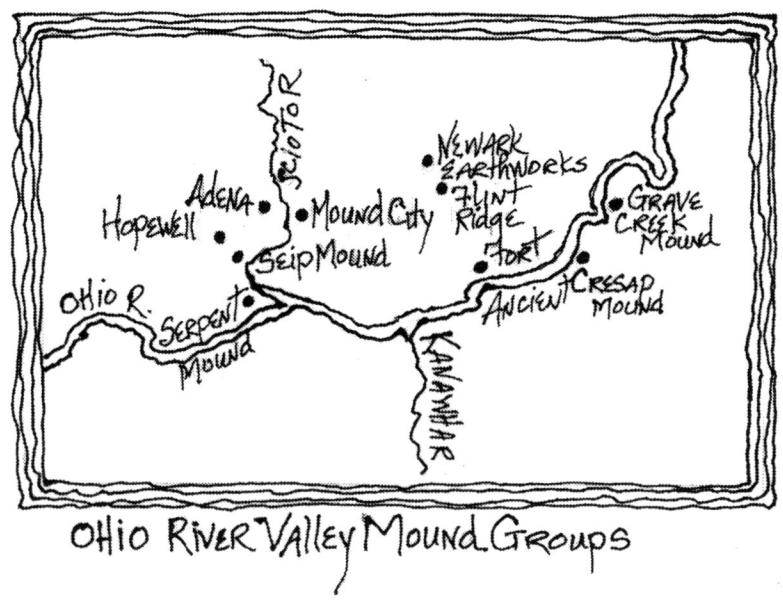

Ohio River Valley Mound Groups Map

Beads

The Mound Builders certainly fancied beads, especially the Hopewellians and Mississippians. One of the favored forms of personal decoration was beads for both men and women throughout the regions. Shell beads were the most commonly worn, made from both fresh- and salt-water shells. Engraved shell gorgets show beads worn in the hair, around the necks, waists, and legs. One adult male burial excavated at Cahokia from Mound 72 revealed a fine leather cape sewn with thousands of shell disc beads.

The Mississippians made beads of shell, copper, stone, clay, and pearls. Bead designs took the forms of casket, thin barrel beads, and finger bone beads, carved to resemble human finger bones. The most highly prized beads were those fashioned from freshwater pearls. Light and lustrous, these were worn in multiple strands to show wealth and favor. The greatest numbers of pearls were found at Spiro, where it was reported that two buckets (galloons) full of pearl beads were found in the Great Chamber A. The pillaging de Soto reported (in the early 1740s) being bribed with quantities of freshwater pearls as tribute, which he accepted and yet preceded to destroy their villages anyway.

5

Adena Mound Building Culture (1000–100 B.C.)

> *A great many mounds were excavated, particularly in those Southern states where flooding caused by the new Tennessee Valley Authority reservoirs would destroy archaeological sites; the WPA activity also concentrated heavily on sites closer to the Ohio mound region.*
>
> —Robert Silverberg, **Mound Builders of Ancient America**, 1968

> Accomplished artists and artisans, the Adena worked with clay and were the first to make grit-tempered pottery, which made their ceramics sturdier. They made tools of stone, shell, antler, and bones. They valued exotic trade goods and adorned themselves with copper bracelets, earrings, and other ornaments as well as shell beads. Adena artisans carved small rectangular stone tablets with snake and bird designs.

The first major mound builders were the Adena people. Perhaps their ancestors were the more southerly Poverty Point People (in Epps, Alabama) and those from Russell Cave (in Bridgeport, Alabama), who built earlier mounds during the late Archaic Period. This began a new American ceremonialism that spread across midland America stretching out to the Atlantic coast. Scientists call them Adena People after excavating a mound on the Adena estate of early Ohio governor, Thomas Worthington, west of Chillicothe, Ohio, in 1901. He once invited the famous Shawnee chief Tecumseh and his brother The Prophet to attend a feast at his home in the early 1800s. The Chillicothe region was the heartland of mound building cultures. Perhaps it is not surprising that Chillicothe became the early historic capital of Ohio.

Adena People, Art, & Villages

Perhaps 500 Adena sites have been studied that were dated to well before the birth of Christ. We do not know who these prehistoric Indians were or what they called themselves, yet their symbols and grave goods indicate that they shared a profound sense of nature and appreciation for life, each other, and death. Studies of Adena burial remains show that these were very tall, robust people. Adena women were six feet tall and the men approached seven feet tall.

The Adena people were obviously talented hunters and fishermen. They hunted deer, wild turkey, and other game with spear and atlatl, and fished with nets, weirs, and hooks for diverse water life. They also gathered fresh-water mussels from nearby streams and rivers and harvested seasonal wild plants—nuts, berries, greens, roots and inner bark. And they were early farmers who cultivated native plants such as sunflower, Jerusalem artichoke, wild rice, goosefoot (wild spinach), groundnuts, and marsh grasses, along with early strains of gourds, tobacco, and corn. Undoubtedly they wore clothes made of hides and furs. They may have also worn some cloth made of fibers spun from milkweed and dogbane, as well as various tough marsh grasses, sedges, and rushes. They used the peeled inner bark of select hickory, oak, hemlock, and maple trees for survival food, as well as to make bark cloth.

Adena craftsmen were famous for their fine platform pipes, especially the Adena pipe, a yellow and red clay effigy made in the form of a man's figure wearing earspools. They made pipes of stone and clay. They fashioned tubular pipes of fine-grained siltstone quarried from natural outcroppings and quarries in the Lower Scioto river valley (in Ohio). Siltstone was excellent for pipemaking as it was easily carved and could be polished to a nice sheen when finished. Adena pipes must have been in great demand and used in trade with distant trading partners near Lake Superior who could supply native copper and with tribes along the Gulf of Mexico who could supply them with exotic shells.

The Adena People lived throughout the Ohio River Valley, where they built small village settlements along streams and tributaries all the way to the Atlantic coast. A typical Adena house was round and from 15 to 45 feet in diameter. The walls were made of paired posts, probably young saplings, tilted outward and joined to support a conical wood roof covered with sheets of tree bark. The house walls may also have been bark covered, and in southern regions may have been wickerwork or cattails and tall rushes or some combination of the seasonally available natural resources.

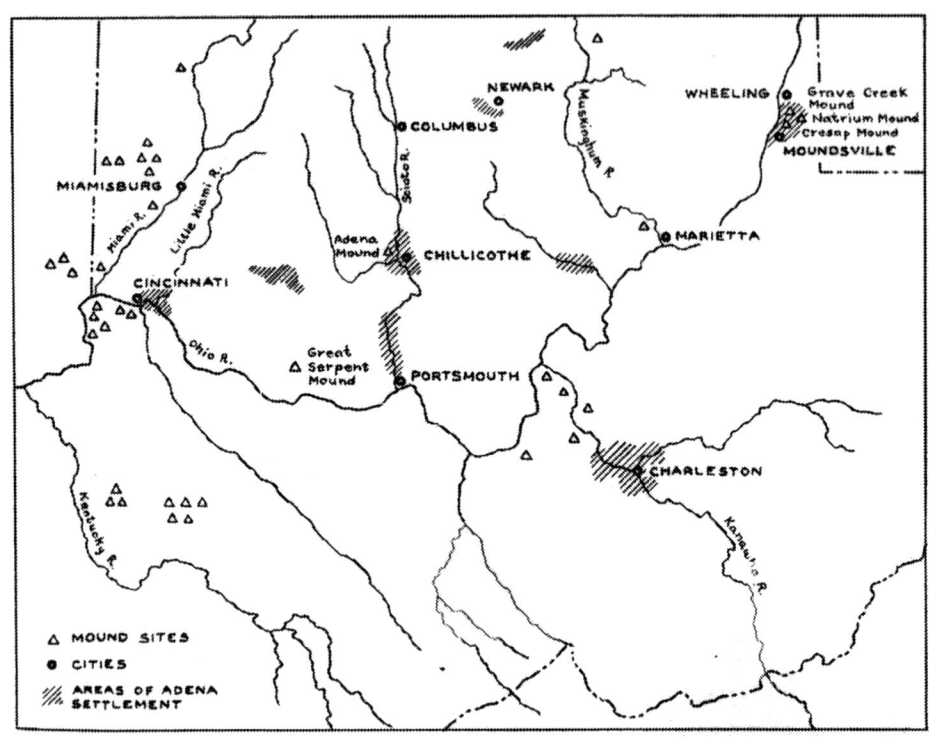

Adena Mounds Sites Map

    The Adena culture area encompassed much of today's Ohio, Indiana, West Virginia, Kentucky, and parts of Pennsylvania and New York. Situated along the Ohio River, the villages were in prime locations between the Mississippi tribes to the west and the tribes to the east. These people developed a network of trading complexes along these central rivers. Each village or group of artisans created unique objects to trade, along with whatever surplus foods and medicines they might offer. Perhaps they traded Adena pottery, pipes, cordage, bark buckets, wild rice, groundnuts, and corn. The copper acquired from Lake Superior they hammered flat and shaped into striking bird and fish effigies. From North Carolina and the Appalachian Mountains they acquired mica, which they cut and shaped into fine symbols such as hands and birds. Mica was highly esteemed for its mirrorlike, sparkling, shiny surface.

    As they traded and acquired valuable items from other Indian villages, their own villages grew and prospered. By about 500 B.C. the Adena People were constructing mound building centers, the most important ones located within a 150-mile radius of today's Chillicothe, Ohio. The Adena had well-organized societies because the construction of the mounds took considerable effort. They built mounds generally ranging in size from 20 to 300 feet in diameter. Large amounts of earth must have been moved by basketloads carried by teams of workers.

Adena Burials

The Adena Indians began building conical mounds around 500 B.C. as special burial places. Historian Otis Rice said these early Americans "built mounds over the remains of chiefs, shamans, priests, and other honored dead." Around many of these burial mounds the Adena built great earthen ridges. We are not certain about the purpose of these ridges other than to glorify some of the special symbolism surrounding the dead and the chosen site. The bird symbol seems to dominate in this culture, perhaps suggesting that there was a sacred bird—possibly the eagle or a traditional thunderbird—that posed as special guardian in life as well as in death.

The early Adena buried their dead simply in shallow bark-lined pits. Sometimes they sprinkled the bodies with red ocher or other colored pigments; then they erected an earthen mound over the spot. Later burials became more elaborate. The Adena cremated the bodies of common folk and placed the remains in small log tombs just beneath the earth's surface. Subsequent dead were interred on top of the first burial stretched out in log tombs. Virtually all of the Adena graves have been destroyed by nature and later settlement. The substantial mounds are our only physical records of their burials.

The Adena buried choice grave goods with their dead—all intended to help the individuals in their duties in the afterlife. At first these grave goods consisted of stone and bone tools, ceramics, and personal ornaments, such as beads and bracelets. Later they included ornaments of hammered copper (cutouts of bird in flight), mica, beads, woven fabrics, and carvings (Adena pipes). Some of these finely crafted objects are now in many museums and private collections around the world.

Because most Adena dead were treated more or less alike, scholars believe that the Adena had an egalitarian society, wherein everyone was treated equally in life. Later Mound Builders were less egalitarian, led by a class of ruling elite. Adena villages may have been fairly small at first with several extended families and related clan members living together in small individual houses. As their societies grew, their villages and burial mounds also grew~many of which remain for us to marvel over today. Many sites were saved when modern local people realized their importance and created state and national parks and museums around them.

Adena Mounds

The early Adena people buried their dead in bark-lined pits, covered with slabs of bark, and piled earth over this to form a mound. Later Adena people buried extended skeletons and cremated dead in log tombs and covered the tombs with earth. As time went by, they added other similar burials atop this—achieving a stacked, layered effect that would rise up into a 30- to 70- foot tall mound. Grave objects were often placed with each burial, such as flints, beads, pipes, and mica ornaments. Adena People built up the huge Grave Creek Mound over a period of several hundred years, from about 250–150 B.C., in successive stages. This massive undertaking required the movement of over 60,000 tons of earth in this one West Virginia complex

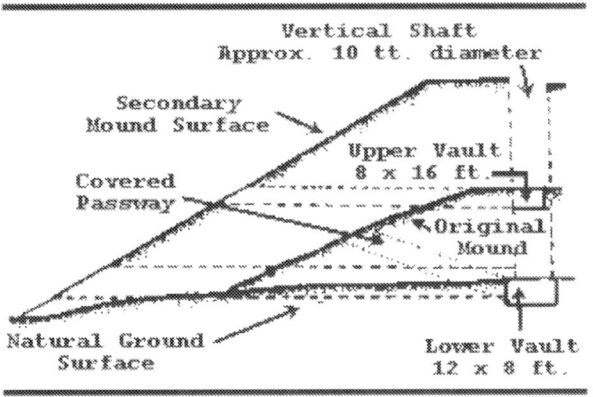

The Adena had well-organized societies since the construction of the mounds took a great deal of effort. The labor of many people was required since large amounts of earth had to be moved by the basket-load. Perhaps for this reason, the mounds were often used more than once. We find in many mounds there are multiple burials at different levels. Over a period of time, the mounds gradually increased in size.

> Further information about the Adena people can be found at the Grave Creek Mound State Park in Moundsville, West Virginia, 304/843-1410.
> www.adena.com/adena/ad/ado1.htm
> www.wvculture.org/history/mounds.html & www.adena.com/adena in West Virginia cultural heritage. Also see www.sunwatch.org/ for the Sun Watch Indian Village/Archaeological Park near Dayton, Ohio

Careful attention was given to construction so that the soil would not easily erode away with wind and rain. Each layer of soil transported to the top of the mound was carefully stamped down and packed into place. Evidence shows that the Adena used different types of sand, clay, and soil in a layering effect in some of their mounds.

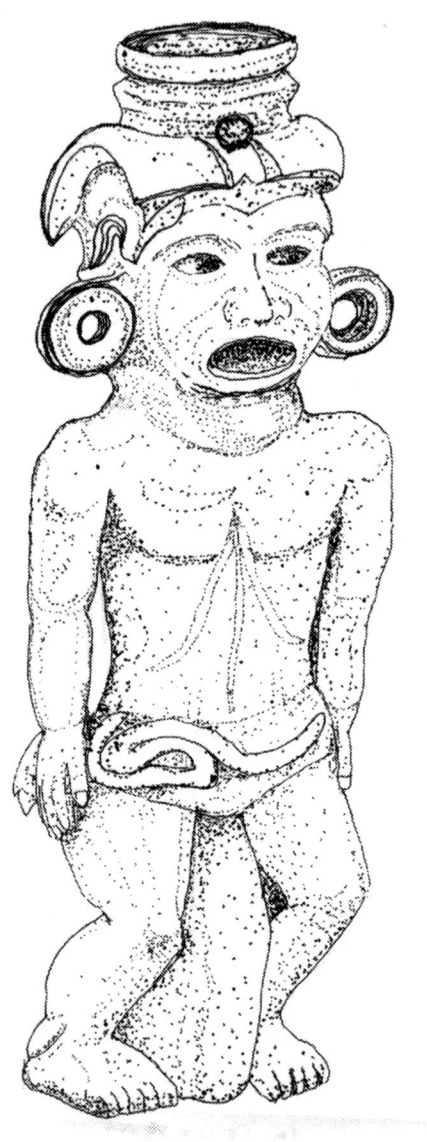

Adena Tubular Male Shaman Pipe–Ceremonial

Perhaps this is a dwarf with a goiter? Such physical differences would have probably been celebrated by the Adena People. Note the large earspools and serpent motif on his breechcloth. This helmeted warrior was fashioned in yellow and red clay. It was found in 1901 in Ohio; now in the Ohio Historical Society, Columbus, Ohio. [It is 8-inches tall and 2000 years old.]

Grave Creek's Mammoth Mound, in Moundsville, West Virginia, is the largest Adena structure. It was built 70 feet tall with a flattened top 60 feet in diameter, measures 240 feet in diameter at the base, and is considered the largest prehistoric Indian burial mound of its kind in the world. It holds about 57,000 tons of earth. Digging and removing so much soil left a sizable ditch surrounding the mound.

According to the Grave Creek museum, *The mound construction probably began with the death of a very important person. There is no way to know who this person was~great warrior, chieftain or religious leader. We know that 25 to 30 years later another important person died and his remains were placed in an 8 by 12 foot vault on top of the mound when it was approximately 35 feet high. The natives then covered this with dirt until the mound reached its maximum height.*

The concept of using each conical earth mound as a cemetery to entomb the bones and finest possessions of dead loved ones was very compelling. Several smaller mounds and earthworks were built nearby Mammoth Mound. These were the revered burial chambers for early Adena dead from encampments and settlements surrounding this site.

Criel Mound in South Charleston, West Virginia, is the largest of about 50 conical mounds on this Adena site created well over 2,000 years ago. These ancient monuments served as burial ground for a whole Indian village located within this area. When the village disappeared is not known, but some scholars suggest it may have remained until early contact times in the 1600s. Criel Mound is 35 feet tall and 175 feet in diameter. It was excavated in 1883-1884 by Professor P.W. Norris of the Smithsonian Institution, who tunneled down from the top and found:

At the depth of 3 feet, in the center of the shaft, some human bones were discovered, doubtless parts of a skeleton said to have been dug up before or at the time of the construction of the judges' stand. At the depth of 4 feet, in a bed of hard earth composed of mixed clay and ashes, were two skeletons, both lying extended on their backs, heads south, and feet near the center of the shaft. Near the heads lay two celts, two stone hoes, one lance head, and two disks.

The archaeologists dug down 31 feet and found many skeletons, including a burial vault of logs containing the remains of eleven Native Americans presumed killed in battle. There was also evidence that some may have been buried alive. According to custom, various weapons and jewelry were placed in the graves.

~~~~~~~~~Take yourself back in time ~~~~~~~~

Imagine an autumn day 2,500 years ago at Graves Creek: the crisp fall weather has turned mild again; two young hunters set off with their atlatls and light spears to hunt deer; they are dressed in soft deerskin tunics, leggings, and breechcloths, with small hunting bags slung across their chests. They wear moccasins made of tough winter elk hide and their long black hair is carefully braided and fastened with partridge feathers. They can hear the women working in the clearing pounding acorns and hickory nuts into meal for soup. Some women are drying and smoking catfish over a fragrant fire. The children are off collecting hazelnuts, sassafras, and spicebush berries in rabbit skin pouches to flavor the families' foods.

Smoke rises from the central cooking fire in the open plaza where the women are working. Four sturdy bark- and rush-covered wigwams surround them tucked back into the woodland edges of the clearing. One man sits in the sunshine pecking stone against stone, creating chert and flint spearpoints. Another man sits nearby weaving strands of peeled hickory bark into sturdy fiber and cordage. Three teenage girls have gone to the creek with elmbark buckets to haul water for cooking and washing.

This small settlement of falcon clan people have lived in this region for seven moons now, and they each work to gather, fish, and hunt for enough food for the approaching winter. They must work harder than usual because the majority of men and boys are nearby working to build the great burial mound higher before the earth freezes. Everyone would like to be there helping, but only the best workers are required to prepare for the seasons of need ahead.

In two more moons they all will gather at the Great Mound to celebrate the "Moon of Frozen Earth." They will dance and perform the ancient rituals to call back the sun. Days of fasting and cleansing will follow. Every home will be swept, cleaned, smudged, and then receive "new fire" for the seasons ahead. The clan leaders have gathered large amounts of cedar leaves and shredded bark to dry before the fires, to be used later for smudging. The smoky fragrance will make everything smell clean. The storytellers will gather families around hearth fires to tell the winter stories

~~~~~~~~~~~~~~~~~~~~~~~~~

> Perhaps such events might have happened something like this long ago in West Virginia. It is difficult to know much more for sure. Archaeology gives us only small remnants of ancient people's lives.

About 2,000 years ago the Adena culture began to slowly give way to a more complex culture as populations increased and Native people evolved. We can only imagine what and where their villages were, yet their amazing monuments~so many carefully constructed burial mounds~stand in mute testimony to Adena achievements.

6

Hopewell Mound Building Culture (200 B.C.–A.D. 500)

o o o o o o o o o o o o o o o o o o o o o o o o o o o o o o

No ancient site anywhere in the world has yielded as many pearls as the Hopewell sites.

—The Field Museum of Natural History, Chicago, Illinois

> Scientists named them Hopewell in the 1890s for a farm owned by Captain Mordecai Hopewell near Chillicothe, Ohio, where more than thirty-eight mounds were found enclosed by an extensive, geometric embankment. This sprawling mound group revealed many details of a sophisticated society of engineers and artists with a keen sense of design.

A new Pan-Indian religion evolved from the Adena culture that scientists would call the Hopewell culture, after another Chillicothe farm, whose mounds were excavated in the early 20th century in Ohio. The Adena culture may have gradually evolved into or given way to the Hopewell culture around 100 B.C. Perhaps the Adena and Hopewell People lived side by side, were one and the same people, and evolved into the next stage of ceremonialism and sophisticated mound building. Perhaps the Hopewell were "next generations" of Adena People. For seven centuries the Hopewell dominated a large region stretching from modern-day Mississippi to Minnesota and from Nebraska to Virginia.

The Hopewell, like their Adena forebears, were not one group of people but a broad network of many different prehistoric Native American groups who traded with one another and who shared common religious practices and a unifying set of symbols and beliefs. They cared deeply about their dead. Artwork in the shape of birds, wolves, fish, snakes, and hands fashioned from mica and hammered in copper often accompanied Hopewell burials. Perhaps these served as guardians to protect loved ones in the afterlife, or were calculated to impress the gods or spirits with the dead person's importance and degree to which he and she was loved. Religious philosophies must have inspired the reverence paid to exalted individuals whose bodies were placed in earthen mounds along with priceless grave goods to serve the spirit in the afterlife. [See cover artwork.]

Hopewell People, Art, & Villages

Hopewellians possessed greater refinement in art and progressed to a higher level of agriculture than the Adena. They created some of the most impressive North American art ever made. Most of the artifacts that remain were used in burials. Others were used as items of trade. And unlike the Adena, they had a strict class system and division of labor. Some scientists believe that the Hopewell society was divided into a series of rank-ordered lineages, ruling families. Certain genetic straits seem apparent in generations of skeletons examined from some mounds, suggesting that the ruling elite may have intermarried with close relatives.

Local chiefs probably controlled each outlying Hopewellian territory, and village groups got together for ceremonies, feasts, trading exchanges, and contests of skill. They lived in small settlements scattered along river valleys near their magnificent earthworks. Larger villages eventually became cities around major earthworks as populations, farming, and trading increased. Hopewell men were successful hunters and fishermen, and gathered many seasonal wild plants.

Hopewell People probably dressed seasonally in soft tanned deerskins and elk skins. Men and boys wore little more than breechcloths gathered securely at the waist with deerskin cords, during the warm spring, summer and fall months. Fringed deerskin vests and shirts and leggings could easily be added as the weather turned cooler. Thicker deerskins with the hair tanned on would be worn as outer mantles during winter snows and coldest times. Elk and buffalo robes, tanned with the hair on, were probably the family's sleeping robes. Women and girls probably wore soft tanned deerskin skirts held at the waist with braided deerskin or rabbitskin thongs or fingerwoven fiber belts, like burden straps. They added deerskin tunics and longer tops during cold weather. Bare feet were preferred during most of the year, along with fine fashioned leather moccasins whenever needed.

Most Hopewellians probably lived in bark-covered wigwams, which resembled large inverted baskets, and oval longhouses, according to archaeological evidence of decayed post molds (stains in the earth left from decayed house posts). These lodges were readily made of bent saplings buried in the ground and the tops bowed over and lashed together. Large sheets of tree bark, most easily slipped from the trees in spring, were lashed to the frame using peeled inner bark. Sometimes an outer frame was mounted over this to fasten it more securely. A smoke hole was fashioned in the center roof, and doorways were fairly small and low. The typical home would have interior sleeping platforms created from sturdy hickory and cedar saplings. These would be covered with fragrant hemlock and pine branches and a large buffalo or deer hide for sleeping. Interior mats were plaited and woven of cattail, bulrushes, and river cane, and could cover doorways or provide inner walls. These mats were often highly decorated with beautiful patterns and natural dyes. Some very fine mats were used to wrap dead loved ones placed for burial in the mounds.

Hopewell Mound Building Culture (200 B.C.–A.D. 500) 65

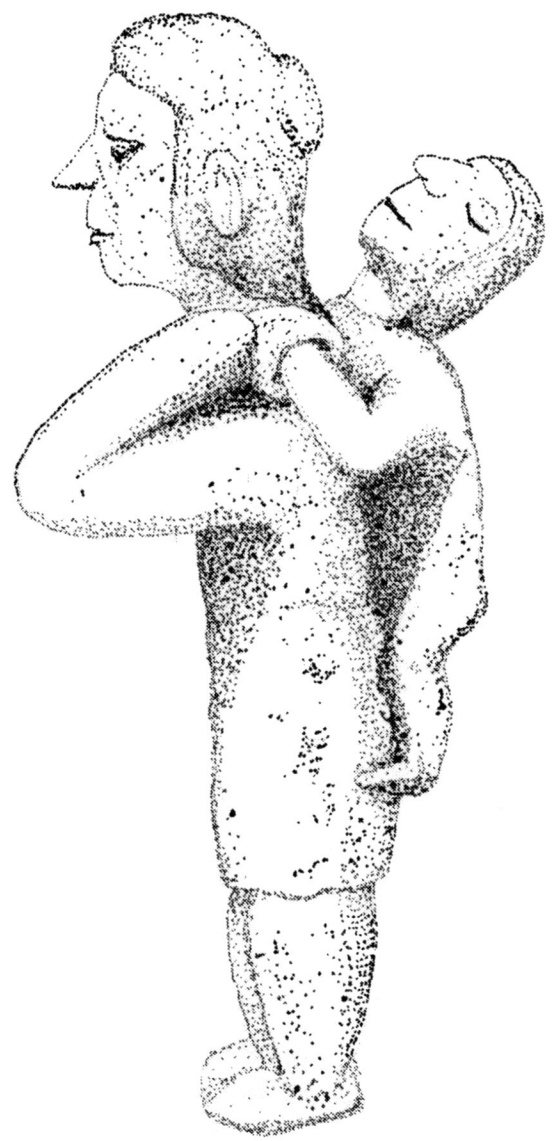

Hopewell Mother carrying Child. This clay figurine was found in Illinois. [Drawn from a photograph; object is in the Milwaukee Public Museum]

Hopewell women were the gardeners who cultivated sunflowers, sunchokes, squash, and some varieties of wild plants rarely grown for food now, such as goosefoot, marsh elder, sumpweed, maygrass, little barley, and knotweed. Men and boys helped to clear the fields. Their division of labor was based upon honoring each other's special skills in societies that were probably matrilineal, wherein the women's lineage and family groups owned the dwellings and the farm fields. This pattern is similar to historic American Indian societies in these regions.

~~~~~~~~ Take yourself back in time ~~~~~~~~

Imagine a midsummer day in Mound City about 2,000 years ago: the river water is warm and clear; young men are spear fishing, while children are capturing frogs and crayfish in the shallow water of a nearby pond. It is warm and no one needs to wear much of anything. Women and young children are scooping mussels from the riverbank to fill large burden baskets. They tell stories while picking raspberries and blueberries along the wooded banks to fill small elmbark berry buckets. One young woman gathers a large cluster of wild mushrooms and wraps them in fresh fern fronds. They walk home passing a potter, who is mixing grit into her clay to strengthen it for shaping large storage bowls. They stop to talk and share some of their fresh harvests with her and her family. Across the path an old man fastens woven mats over bent-pole frames to repair his small summerhouse. They stop again to talk and share some of their fresh harvests with him for his meal.

Dogbane and milkweed stalks are drying in the sun; they will be stripped and their fine, rot-resistant fibers twisted into string to make fabric, burden straps, fishnets, and fishing line. Close to the village a toolmaker sits in the shade pecking antler against flint to create fine spearpoints and arrowheads for the young hunters. His oldest son sits nearby working to fashion a large flint hoe for his mother.

There is an air of excitement throughout the village because the trade chief and his men will be returning by the full moon from their long trip east for valuable supplies. Their canoes will be packed full of marvelous new things: sheets of mirrorlike mica, seashells, and shark teeth, and more tobacco seeds. Everyone is eager for the five days of trading and market festivals that will begin when they return during the Moon of Ripening Corn.

Artisans are making copper and silver pendants, bracelets, and breastplates from the raw materials recently brought back from the Great Lakes, far to the north. The pearl men are opening fresh mussels and sorting them for food, while saving the precious pearls to make into jewelry. They save the best shells to fashion into gorgets, spoons, and scoops, and the broken shells are pounded into grit for the potters. Several men are cutting copper sheets into birds and hands in preparation for the ceremonies planned at the earthworks being constructed on the bluffs above the village. Most of the men and the lynx clan engineers have been moving earth for many weeks to increase the height of the great burial mound since the big spring burial rituals. Another group of men have been building a low outer mound encircling the ceremonial complex. The whole area has a wonderful feeling of power and energy unlike any other place in the region!

> Mound City is now the Hopewell Culture National Historical Park just north of Chillicothe, Ohio in the Scioto River Valley.

Hopewell Trade Networks

Hopewell People developed far-flung trading networks reaching out from their core settlements in the Ohio and Illinois River valleys. They established the first large-scale trade network in precontact North America, expanding greatly upon all that the Adena People had developed before them. The Hopewell influences stretched from Mississippi to Minnesota, and from Nebraska to Virginia. These talented farmers, artists and artisans had much to trade, and they desired exotic materials from far-flung regions.

The Hopewell built a number of different kinds of mounds and earthen works—huge conical burial mounds and many with multiple burials. Extensive embankments enclosed most, perhaps as testament to their sacredness. There was a great concentration of earthworks in Ohio, an old Iroquois word for "beautiful." The Hopewell Culture flourished about the same time that the Parthenon was built on the Acropolis in Athens, Greece (435 B.C.), and Mayan temples were being built in Mexico (300 A.D.).

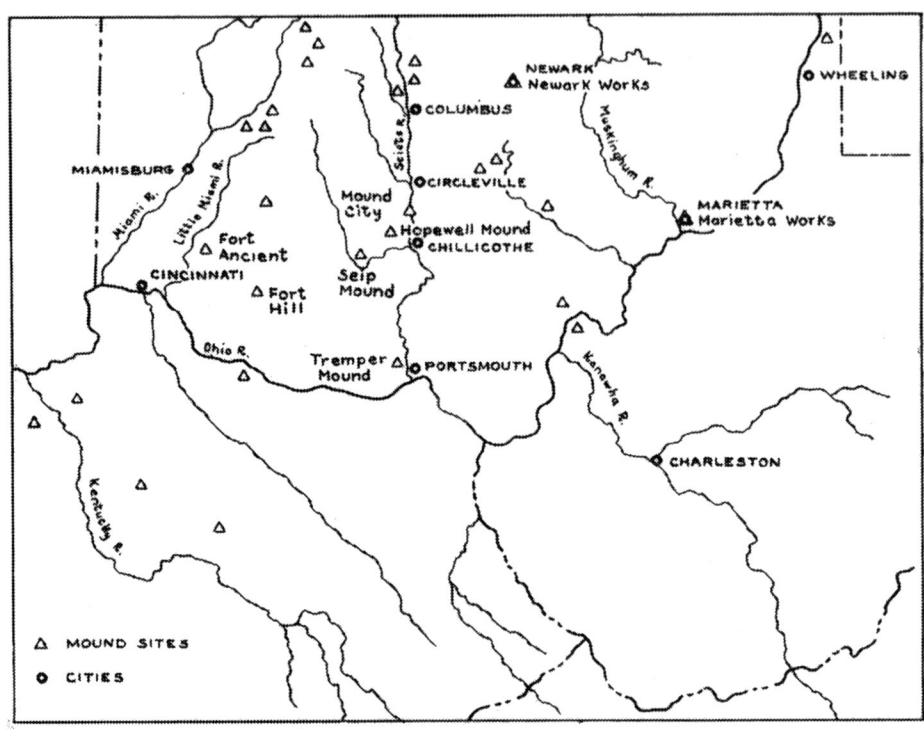

Hopewell Mounds Sites Map

Hopewell Burials

The Hopewell buried their dead with magnificent objects to accompany them in the afterlife. They made these objects from resources they found at home or that they obtained in trade—conch shells and sharks teeth, from the Atlantic and Gulf of Mexico, copper from the Great Lakes regions, silver from Canada, and chert from North Dakota were etched and fashioned into delicate objects. Volcanic glass (obsidian) from the Black Hills and Yellowstone, alligator skulls and teeth from Florida, freshwater pearls from mussels in the Mississippi and Ohio Rivers, and carefully shaped ornaments and symbols made from mica from the Appalachian Mountains filled their gravesites. They are also famous for making platform pipes of pipestone. These have a cylindrical bowl resting on a straight or curved base. Sometimes the bowls are shaped like or depict animals. Some placed their dead in subterranean charnel houses that they then covered over with a great mound of earth. Many gravesites show the effects of burning. Perhaps the

Hopewellian, like other ancient cultures, burned grave offerings in order for the spirits of each object to better follow along with the spirits of the dead into the UpperWorld of the AfterLife.

One mound excavated in 1847 by Squier and Davis in Mound City contained more than 200 carved stone pipes. "The bowls of the pipes are carved in miniature figures of animals, birds, reptiles, etc. All of them are executed with strict fidelity to nature and with exquisite skill." A particularly stunning example is a polished black bird-and-fish effigy pipe depicting a spoonbill sitting atop a catfish. These pipes have been dated to about 100 B.C. Evidence also suggests that a ceremonial structure, like a council house, once stood at the site of each mound within Mound City.

The Hopewellians harvested vast quantities of pearls from freshwater mussels, which they used in trade and to adorn their bodies. Both men and women wore pearl necklaces. Some of their tombs contained several quarts of pearls, and many show the effects of burning. The ancient Greeks and Romans were also using pearls lavishly during this period.

Hopewell Mounds

The Hopewellians continued cremating their dead and surrounding the buried ashes with costly objects to accompany the departed spirits along their journeys into the Afterlife. Some special burials were full form, bodies not flexed or cremated. The burial mounds became long-lasting monuments created by the living to their beloved leaders and their own unique social orders.

Hopewell Mound Group, Hopewell Culture National Park protects a group of Hopewell mounds within a 13-square acre enclosure in Chillicothe, Ohio. (Chillicothe is a Shawnee Indian word for "village" and was possibly also a tribal name.) With 38 conical mounds, most enclosed as an earthen embankment, Mound City preserves one of the greatest centers of Hopewell burial mounds. This was an ancient village and burial site for people living along the Scioto River before A.D. 200.

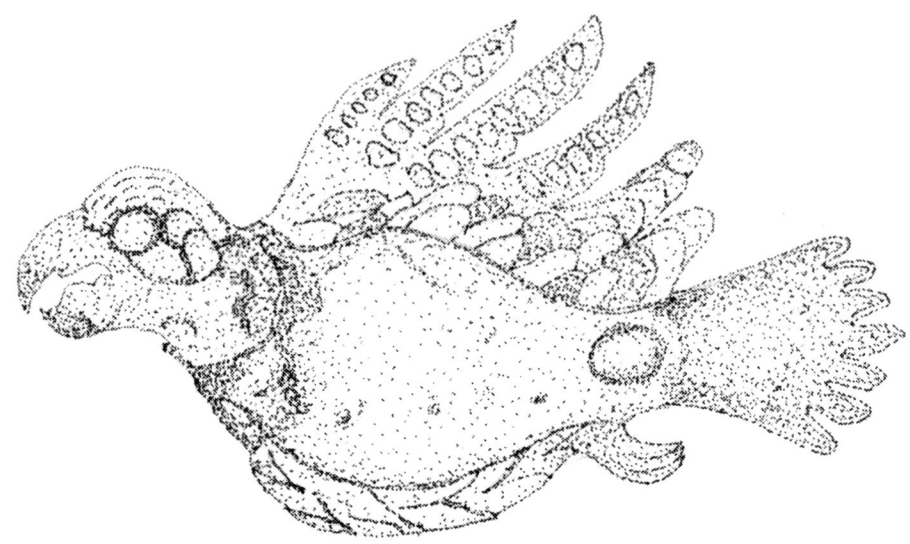

Hopewell Hammered Copper Hawk—found in Ohio Mounds. Symbol of the UpperWorld

Millions of Freshwater Pearls

There are nearly 300 species of freshwater mussels concentrated in eastern North America's rivers and lakes. The mound builders harvested them for food and used their sturdy shells for making utensils, tools, and jewelry, as well as for tempering their pottery. These pearly shells were easily polished to make eating and cooking spoons and small shovels and hoes secured to strong handles.

Enormous numbers of freshwater mussels were harvested to make pearl buttons for clothing in the late 1800s and early 1900s. This became a multimillion-dollar industry in the Midwest until the invention and use of plastics in the 1940s.

Modern environmental impacts were increased when the Japanese used these mussels in their cultured pearl jewelry markets in the 1950s. Thousands of tons of mussel shells were harvested each year and exported to Japan to supply the cultured pearl industry. It is not surprising that our freshwater mussels are one of the most endangered groups of animals in North America. More than 40 mussel species are listed as federally endangered, according to the U.S. Fish and Wildlife Service.

> Today there are thriving pearl (mussel) farms along the Tennessee and other major rivers in the eastern heartland. These modern industries feed growing markets.

The **Newark Earthworks** in Newark, Ohio, is a Hopewell complex that sprawls across four square miles embracing several large compounds connected by causeways. Scientists studying this Hopewell complex believe that there was once a **Great Hopewell Road** that connected Newark mounds to those at Chillicothe Mounds complex 60 miles away. This broad straight road once channeled great activities and communication between these major sites. Unfortunately much of this route has been destroyed by city and urban development.

Today the city of Newark, Ohio, sprawls across most of this area, yet Moundbuilders State Memorial Park preserves what is left of the original complex, including the 1,200-foot diameter grass covered mound. The earthworks in Newark range from 8 to 14 feet tall, encircling three lower mounds that are connected.

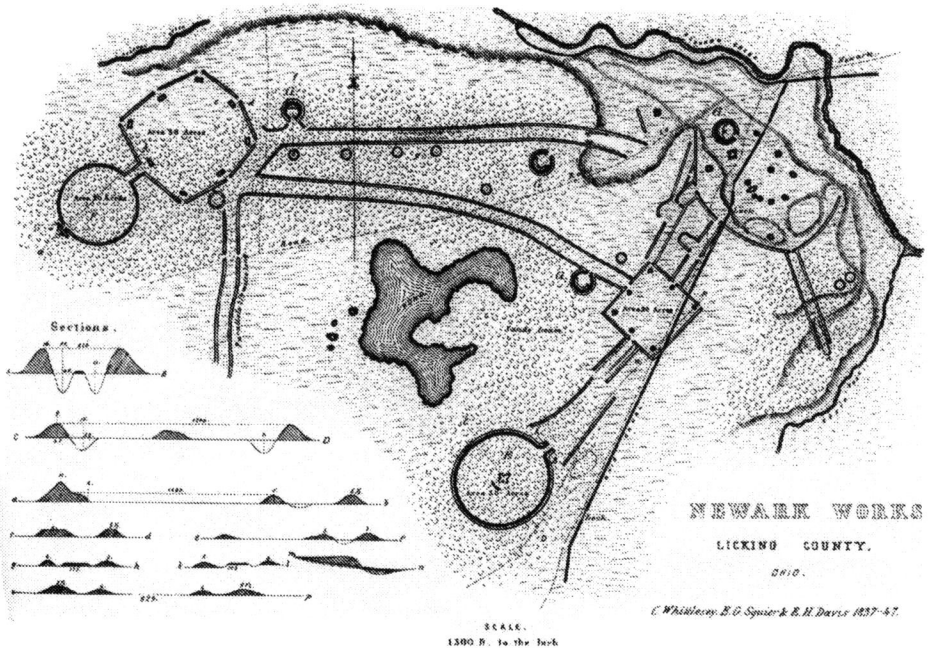

Diagram of the Newark works. Squier and Davis, 1848.

Mica Grave Mound, excavated in 1921, held remains of a wooden building over a shallow clay basin almost six feet square, lined with sheets of mica. This held the cremated remains of four people, along with raven and toad effigy pipes, a copper headpiece, and obsidian tools. Herein, too, were bear and elk teeth, and a cache of 5,000 shell beads, with two copper headdresses. One copper headdress had three pairs of copper antlers, and the other represented a bear, with hinged ears, and legs riveted on. Sixteen more burials lay beneath this. Long after the Hopewell People were gone, other American Indians buried someone in this mound. They were called the Intrusive Mound Culture.

The Mound of Pipes, excavated in 1847, was one of the smaller conical mounds. More than 200 finely carved stone pipes were recovered,... *The bowls of the pipes are carved in miniature figures of animals, birds, reptiles, etc. All of them are executed with strict fidelity to nature and with exquisite skill.* The earliest date for the pipes was 100 B.C. Close connection with the natural world and ceremonialism is echoed here in these finely crafted pipes. Why were they buried all together here? Perhaps this was a monument to a master carver. Pipe smoking was evidently an important ritual.

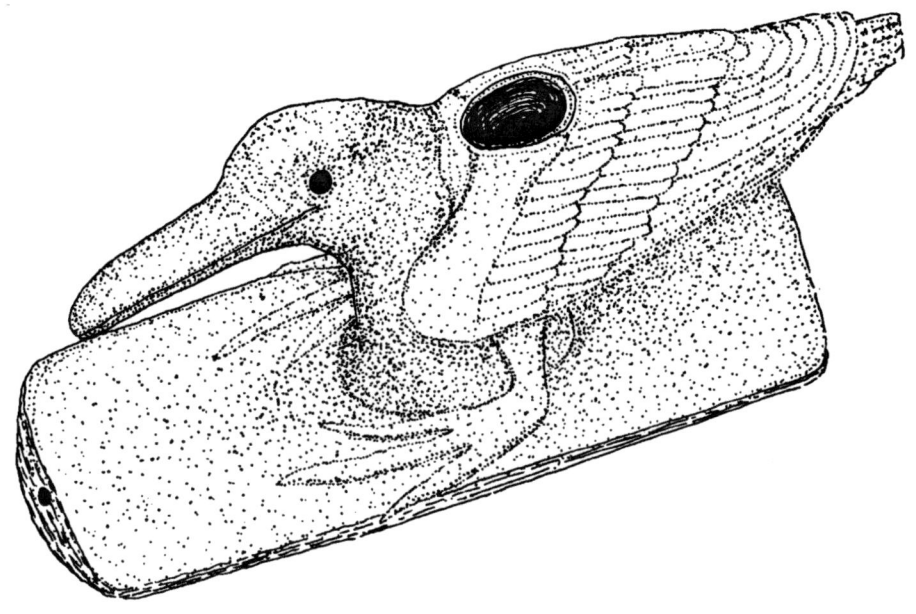

Hopewell Black Crow or Raven Platform Pipe
Mound City, Ohio

Inside **Long Mound** was a circular clay platform holding ashes and cremated human remains. With these were stone and copper objects, pottery fragments, and many spearpoints made of obsidian, garnet, and flint. All was covered with a low earthen mound, which was covered in turn with five alternating layers of sand and earth. A thick layer of gravel and pebbles capped and finished it. Each mound was so different!

Perhaps Mound City's earthworks suggest the social ranking in Hopewell society. Possibly some mound centers were calculated to impress trading partners and visitors from other mound centers. Tremendous planning, engineering, and physical labor were necessary to create every structure. Each mound became an impressive reminder of certain leaders or families for all to see. Yet this is only one Hopewell site and each site is distinctive. So many have been lost to progress and vandalized by "treasure seekers."

~~~~~~~~ Take yourself back in time ~~~~~~~~

Imagine a late summer day in Mound City 2,000 years ago: five men dressed only in breechcloths, their bodies painted with dark clay, occasionally dotted with white, and a red clay symbol of an open hand on their chests; their heads are shaved clean but for a small tuft of long black hair at the back of the crown. Raven and vulture wing feathers are attached to the tufts of hair. These are Vulture Clan Death Priests busy preparing thirteen sacred bodies of the old chief and his wives and warriors for cremation. They work quietly in the Ceremonial Lodge, a charnel house mortuary, atop the Central Mound, much as their fathers and grandfathers had done more than twenty summers ago. This elite Vulture Clan is the only group trained from birth to deal with death, and the rituals of honoring the dead. They know how to prepare bodies after death for cremation and burial. They are highly respected for this.

The weather has turned dark and moody. Thunderstorms are building over the western hills and will bring lightning soon. Perhaps this is auspicious. People have gathered nearby for days of celebrations and feasting preparing for the final rites~releasing the souls of their beloved ones to travel the Great Star Belt.
Inside the large bark covered Longhouse the priests have assembled copper and silver effigies for the old chief; his body is wrapped within the finest cane splint mats, as are the other twelve bodies. They rest side by side on a bark covered log platform three feet above the old mound. Small doorways open to the east and west~the pathway of the spirits.
The bodies have dried out thoroughly and were wrapped with choice fragrant herbs of yarrow, wild mints, skullcap, spicebush, and sweet fern. Now the men prepare a large fire underneath the bodies, as their ancestors had done for an earlier chief thirty summers ago when this mound was first begun. They have been working on hallowed ground and it is time to burn everything~releasing the spirits and their good works to the Fire Gods to be cleansed and carried up into the Sky World.
People begin to stream uphill to the Ceremonial Lodge led by the Falcon Priests. Many people have brought personal offerings of foods and crafts for the final ceremony. Hundreds now gather around the outside of the lodge, singing the death celebration song as they place their offerings beside the lodge. Each Clan Priest gives prayers and a symbolic offering. The crowd backs away and forms a large circle surrounding the burial. Finally the eldest Vulture Death Priest ignites the huge fire within the lodge. He walks out to join the people surrounding the Central Mound. They continue to sing and chant as the whole lodge catches fire and blazes up into the sky. Hours pass.

Sheet lightning moves in overhead, searing the clouds. It is a powerful, unforgetable evening. Night becomes electric with the storm and the death rituals. Rain comes softly at first, then begins pelting down, dowsing the last of the burial fires. The people continue to chant and dance all night around this mound until dawn.

Rain ended hours ago. After brief rest and a morning feast, they will begin immediately to cover this space with fresh earth. Strong winds have picked up after the rains, drying the surface soil. The digging and mounding begins again, and will continue for many days here. Each night there will be drumming and dancing around sacred fires atop the mound tamping down the fresh earth. This will continue for 28 days until the new moon. These are the rituals at important gathering times that renew everyone's spirits. Stories will be told of this time for generations to come.

The Seip Mound State Memorial complex in Bainbridge, Ohio, contains the great central **Hopewell Mound,** which stands 30 feet high, 250 feet long, and 150 feet wide. At least thirty mounds once stood at Seip Earthworks within a large geometric earthwork west of Chillicothe. These mounds represented more civic and ceremonial aspects of Hopewell life, whereas more burials were made in the Chillicothe complex. Several of the large flat topped mounds once supported Hopewell buildings~perhaps temples and living apartments for the priests who ruled over ritual life.

A huge ceremonial platform mound at **Nanih Waiya Historical Site** near Noxapater, Mississippi, is considered the Great Mother in Choctaw creation stories and the ancestral home of the tribe. Hopewell People constructed this mound about A.D. 400. The Choctaw still hold their annual Green Corn Dances here each year. They call this place, this large earth island, "Mother." There is a large burial mound nearby. Because of its sacredness to the Choctaw People, this mound will not be touched (excavated).

~~~~~~~~~~~~~~~~~~~~~~~

Hopewell Bear Skin-Clad Shaman, ca. 400 B.C. known as the Wray figurine.
It was found in an earthwork enclosure at Newark Earthworks in Ohio.

THE CHOCTAW SACRED HOME

> This ancient origin story explains where the Choctaw Indians came from and how they settled where they are today. The Choctaw are considered one of many tribal descendants of the Cahokians. The Choctaw return each year to celebrate their Green Corn dances at the huge platform mound at *Nanih Waiya*, the "Great Mother Mound," in Mississippi. [see appendices: sites.]

The Earth Mother guided the Choctaw to their sacred home in the lower Mississippi Valley. The Choctaw elders chose two men to lead them to the tribe's sacred home. A shaman told the men to cut a young cottonwood tree and strip it clean to make a marking pole. Then he painted it and set it in the ground saying, "By morning the stick will be leaning in the direction that the tribe must travel." They trusted the holy man.

The next morning the painted stick leaned toward the southeast, so the people began their journey in that direction, taking the stick with them. They journeyed on for years, while people died and children were born. They saved their dead and carried the bodies, wrapped in beautiful woven mats, along with the tribal migration. The stick was always planted every evening and its leaning direction followed each new day.

Finally one morning the stick remained upright. As they consulted the stick and each other, they knew they had reached the place where they were meant to settle. The leaders urged families to place dead loved ones around the painted stick and cover them with sheets of cypress and cottonwood bark. Each day the people spent hours hauling river sand and fertile earth to blanket the giant grave with different types of earth from the region. This continued for many months until the Great Mother Mound had grown very tall and fat beside the River Beyond Age. Then they held a large ceremony of celebration and gratitude for many days.

Every year the people return to this sacred site to celebrate for days and Dance to the Green Corn, the first under-ripe sweet corn. This is the time of annual renewal, when elders are honored and babies named, and the ancestors and earth are honored. Many tribes hold Green Corn Celebrations throughout America—each honoring their own unique traditions, while giving thanks to the Earth and the spirits of all growing things. Spirit Plates are set and a small amount of every food and drink is spilled upon the ground to "feed Mother Earth."

The Hopewell culture flourished for more than 500 years and then it, too, began to decline around A.D. 400. Perhaps the people had exhausted their natural resources, or favored a move toward another form of ceremonialism. Perhaps they gradually developed new lifeways and merged into a new group of mound builders, which grew better strains of corn and evolved toward an easier lifestyle. Scientists called the next major culture phase the Mississippians because many of their mounds were created in the Mississippi River Valley. This culture group is also referred to as the Temple Mound Builders because of the huge flat-topped mounds they built to support great temples.

7

Effigy Mound Building Culture (A.D. 350–1300)

> *Upperworld or sky mounds, particularly in the shape of birds, occur throughout the effigy mound region, but are most common in western Wisconsin, especially in the hill country...along the bluffs and terraces of the Wisconsin and Mississippi River valleys.*
>
> —Robert A. Birmingham & Leslie E. Eisenberg, **Indian Mounds of Wisconsin**, 2000

> The Late Woodland Period is known for the effigy mounds. These enigmatic effigy mounds populated a small, special region that includes northeastern Iowa, southwestern Minnesota, and western Wisconsin. Scientists refer to the ancient builders of these artistic landscapes as the Effigy Mound Culture.

Mysterious effigy mounds emerged. Perhaps as many as 10,000 of these awesome and artistic animal-shaped mounds were constructed from A.D 350 to 1300. Ancient Americans shaped huge earthen likenesses of bison, eagles, water birds, lynx, panther, lizards, and turtles. Possibly these were honoring symbols for the society's various clans, giant territorial markers, or hallmarks for sacred healing societies. Archaeologists continue to puzzle over their meanings. When you visit these sacred sites you feel the special energies, while the effigy mounds capture the imagination.

Less than two hundred years after the beginning of the Hopewell period peoples in this region began building the huge earthen monuments in the shapes of animals and also several human forms. Archaeologists call these great works effigy mounds because the individual shapes seem to represent human and natural animal forms.

What was the purpose of the mounds? Some archaeologists believe they were built to honor the dead, and certain clan and creation figures that were considered to be sacred. Others believe they may have served as ceremonial centers where important rituals were held. They certainly seem to add immense energy to the earth.

Effigy Mound People

These talented engineers and laborers were a subgroup of the Mississippians.

They practiced landscape enhancement for almost a thousand years, and their works continue to mystify and dazzle contemporary society.

The Effigy Mound peoples were hunters, gatherers, and fishermen who moved around a great deal. After about AD 1000, they added corn agriculture to their economy. They used the bow and arrow and made other tools from stone, bone, shell, and wood. Their pottery vessels were extraordinarily thin-walled, made with exceptional skill and decorated with complex geometric designs. Their seasonal cycles saw them in family groups scattered over the landscape in winter as they hunted large numbers of deer. They gathered in large, but temporary, villages in spring to catch up on the news, exchange gifts, perform ceremonies and build mounds. Fall was a time to gather nuts, fruits and other wild foods and to get ready for the long winter months. Analyses of their skeletons indicate that they were well-nourished and healthy people.

Around AD 1000, two new groups moved into the area. Archaeologists refer to these as "Oneota" and "Mississippian" peoples. The Oneota were hunters and gatherers like the local residents, but the Oneota grew corn in large quantities and led a more settled life. Mississippian peoples, immigrating from the Cahokia site in Central Illinois, built temples on top of pyramid-shaped, flat-topped mounds and surrounded their permanent villages with massive protective stockades. There are few details about how these three peoples got along with each other, but evidence implies that warfare was common. By AD 1300, Mississippian towns were abandoned. Perhaps Effigy Mound peoples joined the Oneota, since only the latter group lived in the area when the first explorers arrived.

While the Effigy Mound Builders also built conical and other mounds, they did not bury their dead with grave objects. The Effigy Mound Culture lasted seven or eight centuries after the Hopewell disappeared. People in the region leading a farming and hunting life continued to build these mysterious mounds. Then effigy mound building simply stopped.

Effigy Mounds especially populated the regions that would become the state of Wisconsin. The unusual collections and diverse shapes tantalized early settlers, who did not know what to make of them. Wisconsin was covered in more than 15,000 mounds!

Distruction of Effigy Mounds

We were rather irked at the large number of Indian mounds we had to plow down. There must have been at least twenty-five on our land. [Section 36 in Honey Creek Township.] They were particularly numerous in a field close to the creek, near the old village. Some were shaped like animals and some like birds, and all were from three to five feet high...I suppose we should not have destroyed them. But they were then regarded merely as obstacles to cultivation, and everybody plowed them down...Oswald Ragatz, "Memoirs of a Sauk Swiss" Wisconsin History Magazine 19(2) 214; Dec. 1935.

Signals to the Gods?

Perhaps the many builders of the ancient earthen pyramids were reaching toward the Sky World in order to please the Creator, the Great Mystery, and the spirits of their ancestors. The mounds seem to affirm a deep reverence for the earth as well as an appeal to the Sky World and a sense of dominance of the earthen landscapes. Sky Watchers would study the universe from the prominence of temple mounds for signs of changing seasons. Celebrations of the equinoxes and solstices must have been large events in ancient mound builders' societies. Charting the eastern sunrise from the mounds would have held power.

As the mound builders became ever more successful gardeners and farmers, their appeals to sun and moon and the earth's fertility must have grown proportionally. Seasonal ceremonies and rituals were performed to help assure ongoing success. The mounds were certainly huge symbols of success. They were also special offerings of the people's very best work honoring their traditional beliefs, and symbols to the ancestors' spirits. Perhaps their sense of the Spirit World caused mound builders to continually raise great mounds ever higher toward the sky, carrying the bodies of their rulers and ancestors ever closer to the Creator.

~~~~~~~~ Take yourself back in time ~~~~~~~~

Imagine a cool, overcast autumn day in Eagle Town more than a thousand years ago. A respected Elder is walking alone, slowly around the outer rim of the huge Red Horn Giant Mound chanting softly, and brushing fragrant smoke in all directions with his prayer feather. He is singing the story cycle prayers of this great creation figure and honoring the ancestor spirits. He will spend the day singing and feeding the giant mound tobacco and fine cornmeal, and communicating with the ancestors' spirits, in order to help prepare for the tribe's harvest rituals.

He wears a fine soft deerskin tunic daubed with red and black paint to show the exact positions of the major stars in the Sky World for this moon time. His tan deerskin leggings are touched with red stripes above fine beadwork depicting summer medicine plants. His breechcloth is deerskin stitched with tufts of ermine tails. He wears a handsome beaded bandolier bag across his right shoulder, from which he occasionally pulls out more dried bearberry and cedar to add in the smoky smudge, which he carries in a huge old tree fungus bowl. Periodically he pulls a fine otterskin bag open and sprinkles cornmeal over the mound. He keeps this medicine bag tied at his waist. He walks slowly massaging the soft earth with his bare feet as if he's in a trance.

The hundreds of people in Eagle Town depend upon the coming harvest rituals. They've worked tirelessly all year to build this glorious Red Horn Giant Mound ever higher and more robust, patting and firming the rich dark earth holding their prayers and ancestors remains. Teams of men, women, and children worked together with joyful enthusiasm to do this work embracing the mound with dignity. The War Chief, Sky Chief, and Trade Chief also worked along with them, as did the young warriors and Sky Priests (in training.) They will host the many clans from Black Earth this harvest festival because their own man mound is not yet finished.

Everyone will bring their cattail sleeping mats and warm hide robes so they can spread out and sleep each night lying against the side of this great mound. They will each offer prayers and ask for healing dreams. Everyone needs this valuable time to communicate with the ancestors. Each clan will host dream circles each morning before dawn. Two families share a campfire, and many small campfires to keep everyone warm will surround the great mound.

The two villages, along with visiting trading partners, will feast and dance and pray for seven days together renewing Red Horn, the Giant Warrior. The ceremonies will begin on the Full Moon of Ripe Corn, in the great cav~The Womb of Mother Earth~where they will all be reborn as they emerge after a night of prayers beside glowing campfires. This is also the time for renewing commitments to their ancestor's spirits with fine harvest foods and prayers.

More than 20 young couples will be celebrated for starting new lives together, and 28 new babies will receive their names. The Clan Leaders from the Otter, Loon, Rabbit, Beaver, Heron, Wood Duck, and Deer Clans will preside over the feasts. These are the strongest women in the tribe. Everyone will pray for continuing peace and fertility in the land and within the seven clans.

~~~~~~~~~~~~~~~~~~~~~~~~

Farming has destroyed most of the effigy mounds in Eagle Township, Richland County, Wisconsin. Man Mound County Park, near Baraboo, preserves only one (218 feet long) of the four man mounds. Remnants of a second (700 feet long) are found within the township of Black Earth. Both mounds were immense and apparently built as single monuments on the valley floor for ceremonial purposes, possibly ancestor worship. Most of the human effigy mounds depict (shamans) wearing a buffalo-horn headdress, probably representations of Red Horn, the giant warrior hero at the center of story cycles told by the related Ho-Chunk (Winnebago) and Ioway Tribes.

Blueberry Shrub, *Vaccinium angustifolium*--valuable perennial foods, beverages, medicines & dentifrices were available from the whole plant.

Wisconsin's Cave of Wonders[1]

Enter the Dream Lodge, access ancient wisdom,
journey into the Sacred Mystery, much like it
was when birch bark torches once lit prehistoric
passageways into three deep chambers.
Ritual ceremonial life unfolded here
beneath 1,300-year-old charcoal drawings
absorbed into native stone holding birds,
pregnant deer, bow-hunters dancing suspended
along cool rough walls of Earth's womb.
Vision quests tied closely to Nature add
Spirit prayers and offerings reflected in
this ancient rock art symbolizing origin myths,
the giant Red Horn, warrior hero,
ritual dances, feast times, family and clan lines.

Infants in cradleboards float across the
ceiling of the cave's entrance; evidence of
ritual headflattening tied to the 1,500-year-old
Indian Mounds nearby in western Wisconsin.
Charcoal line connects to Thunderbird,
perhaps an early "naming ceremony" commemorated
descendants of Ancient Effigy Mound Builders.
They recorded events in their lives with
pigments and carvings on pottery and stone.
Shamans danced journeying to the sources of
game, causes of illness, mysteries of healing.
Return to be reborn here in ritual ceremonies.
Who can decipher these meanings in rock art
created like paper and ink in our society, yet
will our thoughts ever last this long?
This "Driftless Area" lay free of
glacial change ranging across
deep valleys in a rugged landscape;
Ho-Chunk and Ioway homelands today embrace
ancient Earth wombs with countless stories to tell,
ancestors to worship, prayers to echo.

---

1. This poem was published in an earlier form in *ANCESTRAL THREADS: Weaving Remembrance in Poetry & Essays & Family Folklore,* 2003, Iuniverse.

The Gottschall Rockshelter is a cave in a remote upland valley in Iowa County scarcely 10 miles from Eagle Township's effigy mound group. It holds the painted figures telling the story of Red Horn, placed on the cave walls 1000 years ago. The Eagle Township Effigy Mounds Group also holds the largest bird effigy mound ever recorded in Wisconsin.

Etowah and **Ocmulgee** in Georgia were impressive Mississippian villages and ceremonial centers. Distinctive architecture and crafts found at these sites suggest that major chiefdoms ruled and guided extensive commerce from far-flung trading networks. Great Temple Mound once supported three separate wooden structures, and nearby Funeral Mound held seven levels of Mississippians' elite society. These were probably the early ancestors of Cherokee and Catawba Indians, as well as other tribal groups.

In nearby Eatonton, Georgia, rests a most unusual Rock Eagle Effigy Mound formation created entirely of white quartz boulders and cobbles that were carried to this site from some distance. Ancient ancestors of the Mississippians constructed it perhaps 5,000 years ago. Was it a signal to the gods that the people respected their traditions? This impressive effigy has a 120-foot wingspan and seems to also favor a wild turkey or vulture in form.

Many medicinal plants surround this site, creating the impression that this place might have served as an important healing center. Rock Eagle Mound stands alone and was constructed 2,500 years earlier than the Great Pyramid of Egypt.

Etowah is a Cherokee-Creek Indian word meaning, "high tower village" and *Ocmulgee* is a Hitchiti tribal name, meaning, "where water bubbles up"—indicating a spring. These ancient words bring us glimpses of their earlier places of importance. Perhaps these words originated among the ancient mound builders.

Rock Eagle Stone Effigy Mound in Eatonton, Georgia; 5000 years old

Giant Animal Figures

Mysterious effigy mounds were created by more northern groups of mound builders by the thousands; awesome and artistic human- and animal-shaped mounds. Perhaps these were honoring symbols for the society's various clans, giant territorial markers, or hallmarks for sacred healing societies. The Late Woodland and Late Mississippian Period is known for the effigy mounds. They practiced landscape enhancement for almost a thousand years, and their works continue to mystify and dazzle contemporary society. Unfortunately, intensive agriculture and urban expansion has erased most of them.

There was also a fascinating cultural shift in these regions away from sending the dead off with costly grave goods in burials, to shaping effigy mounds to honor the dead, as well as the living. Native copper, flint, and obsidian were still shaped into valuable weapons, tools, and ceremonial items, but kept for the living rather than placed into burials. Most effigy mounds contain no bones, ashes, or grave goods.

More than 200 mounds populate a 1,475-acre site in northeastern Iowa called **Effigy Mounds National Monument.** The mounds were built during the past 2,500 years. Three huge birds and a string of ten Marching Bears command this woodland landscape. These 13 effigy mounds are nearly evenly spaced along an east-west ridgeline. There are 26 animal effigy mounds rising here. **Great Bear Mound** is 137 feet long and 70 feet across the shoulders. Along with these distinctive effigy mounds are several conical and linear mounds containing human burials. There must have been tremendous significance and pleasure in actually building each mound. In spite of the hard labor, the prize was perhaps in doing the work, as well as experiencing the end results.

One of the best-preserved mound groups is at **Lizard Mound County Park** in Wisconsin. Thirty mounds survive today of the 60 effigy mounds once created here. Most of these mounds are only about four feet high now, yet their shapes are amazingly crisp. These gigantic artistic symbols contain no burials or artifacts. They rise only three to four feet above the prairie. Walking among them in deep snow one still sees the fine outlines and forms, a compliment to the original makers. Several large panthers and two birds dominate this site along with many cigarlike and conical shapes.

Perhaps the Animal Effigies simply defined sacred ground, and called people together for ceremonies and rituals of renewal and healing. Possibly these special mounds were shamans' destinations, where they would journey to the nature spirits seeking wisdom and healing. There must be some connections to the cycles of sun, moon, and constellations, and traditional Native American cosmology. Some say the Great Marching Bears represent the Big Dipper circling Polaris, the North Star. Perhaps it was the building of the giant animal mounds—alone—that was the supreme connection. Whatever their origins, they remain silent witnesses of a vanished culture and mysteries.

The many effigy mounds suggest an even closer connection with the spirits of the species represented. Huge earthen snake, bear, eagle, lizard, and bird mounds provided supernatural support for each creature in the natural environment, as well as awesome clan symbols. These earth islands were special sanctuaries and places for honoring rituals.

> The Wisconsin Dells still hold examples of the ancient Effigy Mound Builders. One example is at the Kingsley Bend Wayside on Highway 16 where there are a group of about 20 burial and effigy mounds. There are conical and linear mounds as well as effigies of two 100-foot long bears, a panther with a tail as long as a football field and an eagle with a 200-foot wingspan.

Scattered about the **Beloit College campus** are 23 conical, linear and animal effigy mounds built between about AD 700 and 1200. One, in the form of a turtle, has inspired the symbol of the College. These mounds and others like them are found in Southern Wisconsin and adjacent portions of the surrounding states. Effigy Mound people built them, and current research suggests that they were the ancestors of the modern Ho-Chunk Nation.

The mounds are rarely higher than five feet and usually small, although a newly discovered bird effigy has a wingspan of 1310 feet. They were usually built along bluff tops adjacent to rivers, and the Beloit College group illustrates this pattern. Mounds also occur in groups of from two to more than two hundred. Maps made by early explorers estimated that more than 20,000 mounds existed prior to the arrival of the pioneers in Wisconsin, yet fewer than 3000 remain.

New waves of American settlers pushing west two hundred years ago often failed to understand the importance of these magnificent structures. Unfortunately, intensive agriculture and urban expansion has erased many of them.

The mounds served as final resting places for recently deceased members of the family. Some human remains have been found in many mounds, yet others contain no skeletons. Perhaps they were built as family unity projects, where everyone in an extended family lent their hands to the work as a way of reinforcing their identity. The mounds also served to define territories of such families and the shapes may represent clan totems. Religious ceremonies must have accompanied the building of the mounds.

> All of the mounds on the Beloit College campus have been excavated and reconstructed. Materials found in them, such as scattered fragments of pottery vessels and small pieces of broken stone tools, are in the permanent collection on the College's Logan Museum of Anthropology.

Stretched mysteriously along a bluff overlooking Ohio's Brush Creek is the great **Serpent Mound**, more than one-quarter mile long and forever slithering northward. At its head is an oval mound. Some observers think this represents an egg, which the earthen snake is trying to swallow. Or perhaps the oval mound represents the creature's heart, or an enlarged eye, or the open mouth turning outward to devour anyone who looks at it.

Serpent Mound is one of the most unique effigy mounds in the world. It contained no burials or artifacts within it to tell us more details about its origins or its precise dates. Some authorities reckon that later Mississippian People constructed it over a thousand years ago.

The meanings and purpose of this impressive effigy mound with its coiled tale and long sinuous body has always puzzled observers. Serpents figure prominently throughout Native American lore and usually symbolize renewal, respect, cunning, and mystery. Ohio Indians, however, suggest that the great earthen snake symbolized the Big and Little Dipper, Ursa Major and Ursa Minor, and especially the latter's movement around the North Star. Ohio Indian elders saw the Serpent Mound as a symbol of Ursa Major, the Great Bear, or the Big Dipper.

Effigy Mound Building Culture (A.D. 350–1300) 91

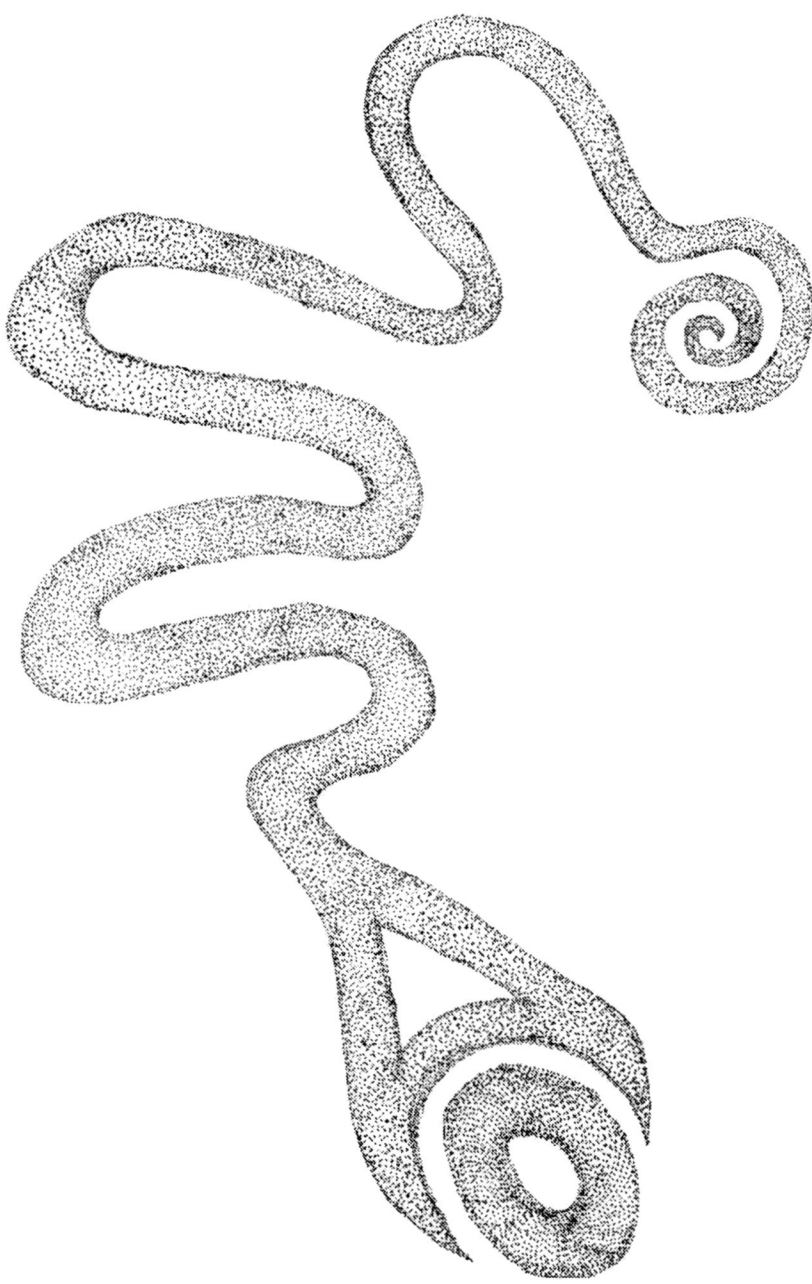

The Great Serpent Mound in south-central Ohio

Later astronomers have suggested that perhaps it was meant to symbolize Halley's comet or a shooting star. Some scholars have found key celestial alignments in the curves of Serpent Mound. Archaeologists carbon dated some charcoal fragments found beneath the earth near Serpent Mound, and the resulting date was about A.D. 1070. This falls between two major celestial events. The intense light from a supernova that produced the Crab nebula first reached earth in 1054. This remained visible during the daytime for perhaps two weeks. Then the brightest showing from Halley's comet ever visible reached earth in 1066.

Did ancient mound builders immortalize these events in the body and symbolism of the Serpent Mound? Is the oval "egg" actually the comet or supernova, and the slithering serpent body the comet's fiery tail? We will never know for sure, yet we will always feel the magnetic appeal of this ancient earthwork.

They lived high on a hill.
They were people who
Were up before first light.
They did not have peculiar customs.
They would not bother anybody.

They were of the following minds~
    Crows, hawks, horned owls,
    Field larks, hummingbirds,
    Blue birds, chickadees,
    Quail, woodpeckers,
    Yellow hammer, whippoorwill.

They did not work, they lived easy.
The best of them were wiser
Than other clans. They did not
Depend on anyone but themselves.
They were the Bird Clans.

—Chickasaw song

8

The Temple Mound Builders: Mississippian Culture (A.D. 750–1500)

> *I wonder if the ground is listening to what is said?... The ground says, It is the Great Spirit that placed me here.... The Great Spirit directs me...*
>
> —*Young Chief, a Cayuse leader*

> Temple Mound Builders (the Mississippians) flourished in America's heartland and spread out across the regions we now call Oklahoma, Wisconsin, and Florida. These dynamic cultures left mysterious objects of fine beauty that continue to impress our modern culture.

Eastern Woodland Indians gradually evolved to even more sophisticated social groups. These were hundreds of individual societies that thrived in villages throughout the Tennessee, Cumberland, and Mississippi river valleys. Scholars called them the Mississippian Culture after the mighty river that was the center of their expansive life.

The Mississippians flourished and built incredible mounds. They were master farmers and tended fields of corn (maize), squash, pumpkins, beans, and tobacco. They lived in thatch-covered homes built in clustered settlements near the fertile floodplains of major rivers. The large flat-topped pyramids that they built reached from 18 to 60 feet tall. Temples were built on the broad flattened tops. Temple Mound Builders expressed themselves in many ways. Their sense of beauty was reflected in the stone carvings and pottery objects they created, and the countless ritual items they buried with their dead.

Temple Mound People

The Mississippians were mathematicians and engineers, artists and fine craftsmen. They were men and women of vision, whose Earth-based wisdom enabled them to prosper. They lived in more structured societies and developed greater trading networks that took them far and wide. Some Meso-American influences are seen in Mississippian societies, from whom they must have gained corn, squash, beans, tobacco, and many ceremonial ideas. As their populations grew in more settled societies, these amazing mound builders specialized and developed amazing talents for personal adornment and artistic expressions in pottery, stone, bone, metal, shell, and especially earth—on a majestic scale.

Cahokian Woman: smoothly polished, carved of reddish brown bauxite stone, this woman is kneeling on a coiled snake. She holds a hoe in her right hand and is stroking or scratching the snake's back. The back of the sculpture shows the snake's body divides into two gourd vines. Gourds were one of the first plants domesticated by early farmers in North America during the late Archaic Period (ca 3000 to 500 B.C.). This is obviously a fertility symbol associated with fortuitous farming. Here is the famous Birger Figurine, about 8 inches tall and dated at Cahokia's peak, between 1000 to 1250 A.D.; the Illinois Archaeological Survey, University of Illinois. [Top of the head has been broken away.]

The Temple Mound Builders: Mississippian Culture (A.D. 750–1500)

Mississippian hunters acquired seasonal game with bow and arrows and hunted many of the same quarries as earlier Hopewell People. The climate was very similar to today's, yet the environment was still considerably more forested and naturally teeming with life. Farming was a greater force in Mississippian life, allowing people to settle and prosper in permanent villages and palisaded hamlets near extensive farm fields that were worked principally by the women and children. Men were required to do the hunting, fishing, trading, and to build greater mounds. Mississippians worshiped a fire-sun deity for whom they constructed breathtaking ceremonial centers. An aristocracy adept at leading great rituals and ceremonies to honor the Sun God and the fertility of Mother Earth governed them. Many of the eastern woodland tribes and Indians in other regions did not participate in Mississippian culture, although some were their trading partners.

The greatest center of Mississippian culture was Cahokia, "City of the Sun." Here, in what is now East St. Louis and Collinsville, Illinois, Cahokians built the largest pre-Columbian city in North America, which was home to perhaps 20,000 people. Cahokians flourished for more than 800 years, from A.D. 700 to 1500, and Cahokia eventually covered much of the rich floodplain where the Mississippi and Missouri Rivers join together. Cahokia occupied about 4000 acres and contained more than 120 mounds. At its zenith, from 1100 to 1200, this city sprawled over nearly six square miles. This became the hub of trade networks all across North America. Meso-American influences are seen in Cahokia that seem to parallel the temples and plazas of the Mayan and Aztec People, who must have been southern trading partners.

Cahokia is the name of a sub-tribe of the Illini Indians, who camped and hunted in this region, and were encountered by French explorers in the late 1600s. We do not know what these ancient Temple Mound Builders actually called this center. Their houses were arranged in rows around open plazas. Agricultural fields lay outside the city, where Cahokians grew corn, squash, and beans, along with sunflowers, pigweed, Jerusalem Artichokes, and lambsquarters as the principal crops. They also managed great stands of wild rice, marshelder, maygrass, tobacco, wild onions and garlic, knotweed, and little barley, while hunting, fishing, and gathering other wild foods.

~~~~~~~~ Take a step back in time ~~~~~~~~

Imagine a winter day in Cahokia a thousand years ago: the air is brittle with cold and falling snow, blanketing several inches of old snow crusting the ground. This is the Moon of Frozen Earth when the Trees Crack, and village life is brisk and playful. Children dressed in warm soft deerskin tunics and leggings, breechcloths, and rabbit fur capes are gathering kindling and firewood for each wigwam in the village. A team of ten hunters set off after sunrise to check their beaver and rabbit traps, and lay new snares for wild turkeys. Women are busily working to tan elk and deer hides stretched on large open frames near warm campfires. They chatter together as they work. Two young women have new babies bundled into cradleboards on their backs. They sing to their infants as they work.

This portion of the settlement here above the Mississippi is populated by the Corn Clan People, who have lived on this floodplain region for more than 20 moons. They are related to the great Trade Chief, who has gone downriver with many men and some women to trade with the Eagle Bluff People seven days south of here. Some of the young women may stay and marry with these gifted trading partners. They have been intermarrying with selected trading clans for as long as one can remember, because it is forbidden to marry within one's own clan. Also the Great Chief wants to keep solid relations with all our regional trading partners, so some of our prettiest young women are often selected to go along on less dangerous travels. They have packed many canoes full with autumn furs, dried corn, squash, pumpkins, beans, wild rice, and tobacco. The Trade Chief also selected many of our finest effigy pipes, and a big medley of dried medicine roots and kinnikinnik from the herbalists and healers. These items are packed into a few of our finest leather and fingerwoven medicine bags. Our clan is famous for their fine medicine bags.

~~~~~~~~~~~~~~~~~~~~~~~~

The Temple Mound Builders: Mississippian Culture (A.D. 750–1500) 97

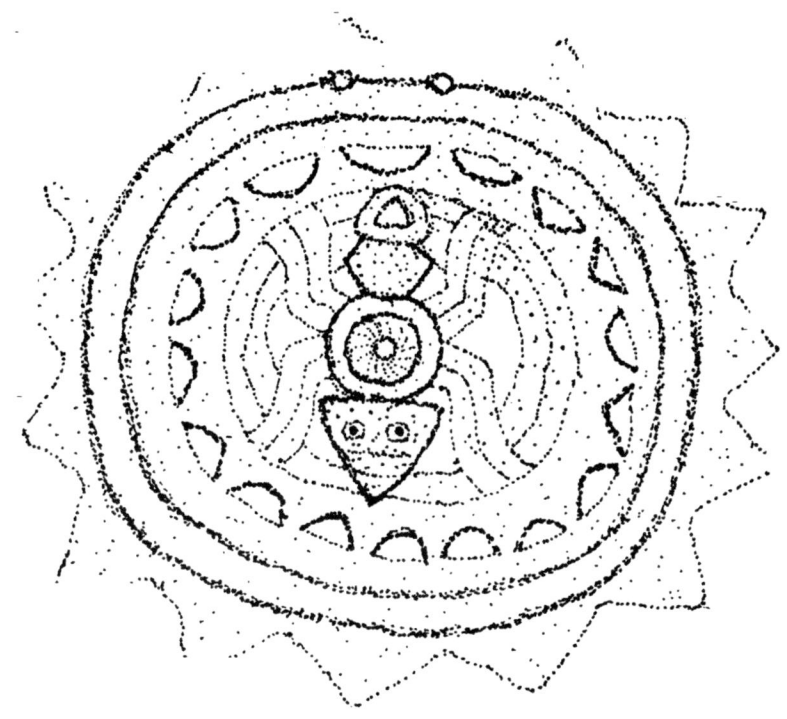

Spider Design Shell Gorget—Southeast Ceremonial Complex
Illinois Mounds, ca. 1000 years old

All over the sky a sacred voice is calling you.
—Black Elk, Lakota Sioux Holy Man, 1873

Mississippian Mounds

Hundreds of Native American societies thrived throughout the Mississippi, Cumberland, and Tennessee River valleys from about 750 until the 1500s. Hunters with bow and arrows and farmers increased food productivity for the villages yielding surpluses, in most years, enough to trade. Shell-tempered pottery became more elaborate in shape and design, as did the immense mounds. People began to construct huge flat-topped mounds and town plazas nearby, where the Mississippian elite ruled over breathtaking ceremonial centers. Yet most of the Eastern Woodlands Indians in other regions did not embrace the Mississippian culture, but continued to evolve within the Woodland culture traditions.

A Native aristocracy presided over the largest and most powerful center north of Mexico~Cahokia, which covered almost six square miles. More than 120 diverse mounds once peppered this majestic site. This was the center of the Temple Mound Builders. The Mississippians carried mound building to whole new heights~literally~while creating ceremonial centers. Huge mounds were being built with flattened tops surmounted by temples and lodgings for the societies' priests. We call these Mississippians Temple Mound Builders because of this major shift in mound building and ceremonial practices. Even more exotic grave goods were buried with the dead.

People began stockading large villages and mound complexes for protection in the Middle Mississippian Period, around 1150. Cahokia was protected by a massive wooden stockade up to 15 feet high encircling a two-mile section of the City of the Sun with perhaps 15,000 logs. Diverse farmsteads and small outlying villages surrounded these large palisaded complexes.

Scholars suggest that the Cahokian people belonged to a group of tribes known as the Dhegihan Sioux, which included the Omaha, Osage, Kansa, Ponca, Arkansa, and Quapaw People. This is based upon some oral traditions and fragments of historical work. Many of these tribes and more may be the living descendants of Cahokia.

Mississippians constructed three distinctive types of mounds in their great city. Large flat-topped mounds supported ceremonial buildings and residences of the elite, the leaders. These are the most common mounds. The Mississippians also built conical and a number of ridgetop mounds, used for burials of important people and to mark important locations. (Cahokians built many more ceremonial mounds than burial mounds.)

The largest prehistoric earthen construction in North America is **Monks Mound**, the great platform mound at Cahokia. It was built in several stages between 900 and 1200, holding approximately 22 million cubic feet of earth. Monks Mound covers more than 14 acres, rising in four terraces to about 100 feet high. Atop it once stood a massive building almost 50 feet tall and 105 feet long by 48 feet wide. The principal ruler must have lived here with his advisors and family while governing the city and conducting ceremonies. This also provided the finest observatory—overlooking all of the surrounding territory.

Monks Mound was named in the early 1800s for the French Trappist monks who lived nearby and once grew their vegetables on the terraces. We can only imagine what its true ceremonial name might have been; possibly Temple to the Sun, because the Cahokians seemed to revere the sun and fire. Indians eventually

The Temple Mound Builders: Mississippian Culture (A.D. 750–1500)

moved more than 50 million cubic feet of earth at Cahokia to construct the vast mounds.

> From the summit of **Monks Mound**, chiefs once looked across the broad central plaza to the Twin Mounds. Deceased nobles were prepared for burial in the temple atop the flat-topped mound for interment in the adjacent conical mound. Temples, homes of the elite, and burials were associated with more than 100 other manmade mounds. Houses with pole walls and grass-thatched roofs were clustered within and beyond the 15-foot-high stockade wall.
> —Cahokia Mounds site, Illinois Historic Preservation Agency

A small ridgetop mound, **Mound 72**, was excavated in Cahokia, revealing nearly 300 ceremonial and sacrificial burials of young women in mass graves. They appeared to accompany a male ruler, the main burial, who was about 45 years old. He was placed on a blanket of over 20,000 marine shells and disk beads, along with a large cache of grave offerings and the remains of others sacrificed to serve him in the next life. The skeletons of four men with their hands and heads missing rested near the largest sacrificial pit, which held skeletons of 53 women ranging from 15 to 25 years old.

Some of the conical mounds, usually burial mounds, were unwittingly destroyed during centuries of farming, road building, and settlements. It is impossible to know how much information we have lost about these dynamic cultures.

"Most Cahokians were probably buried in cemeteries, not in mounds," says the Illinois Historic Preservation Agency. The common folk were buried outside of the central city, probably near their villages in smaller conical mounds, or just below the earth surface. This great culture slowly declined, before European contact, from diseases, internal conflicts, and exhausted natural resources.

> **Cahokia Mounds** preserves the remains of the central section of the largest prehistoric Indian city north of Mexico. Covering nearly 4,000 acres, the Cahokia site was first inhabited around 700 and grew to a population of nearly 20,000 by 1100 Sixty-eight of the original 120 entirely earthen mounds are preserved within the historic area. At the center is Monks Mound, which at 100 feet tall, is the largest prehistoric earthen mound in the New World.

A gradual decline in Cahokia's populations began after 1200. War, social unrest, disease, and declining political and economic powers may have taken a toll. A climate change after 1200 probably affected plant and animal resources and crop production. Years of drought may have devastated the entire region. There was a prolonged period of drier, cooler weather. Depletion of natural resources, especially trees and game animals, as well as soil depletion, contributed to the city's loss of vigor. Cahokia was abandoned by 1400. The ultimate fate of prehistoric Cahokia is unknown. But it is certain that the ancient Cahokians were ancestors of the Natchez, Choctaw, Shawnee, Illini, and Creek Indians who spread prosperous villages across many of these regions.

Moundville, overlooking the Black Warrior River in Moundville, Alabama, was another large ceremonial center sprawling across 300 acres. Between 1000 and 1500 the Temple Mound Builders constructed 24 square and oval mounds here crowned with temples and council houses. Graded paths climb to the mound summits. Here, too, was a stout palisade protecting the several thousand people who once lived at this center. Excavations at this site during the past 100 years have uncovered more than 3,000 human burials. Each mound contained the burials of a few high-status adults along with copper axes and gorgets, stone disks, and various paints and minerals, like mica and galena. A stunning design motif appearing here is the eagle being, a dancing priest. Yet over half of the Moundville graves were for commoners, buried without any grave goods.

Moundville was part of a major symbolic network called the Southeastern Ceremonial Complex, or The Southern Cult. This great triangle stretched from here to **Etowah** (in Georgia), and out to **Spiro** (in Oklahoma), and flourished from 1150 until 1350. A higher degree of social exchanges, trade, and symbolism is found concentrated at these particular centers within the Mississippian world. The Southern Cult symbols are distinctive crosses, sun circles, the forked eye, the hand and eye, and bi-lobed arrows along with falcon, serpent, woodpecker, raccoon, and others. The religious practices recognized the vitality of seasonal cycles, which must have honored planting and harvesting festivals, along with reverence for life and death.

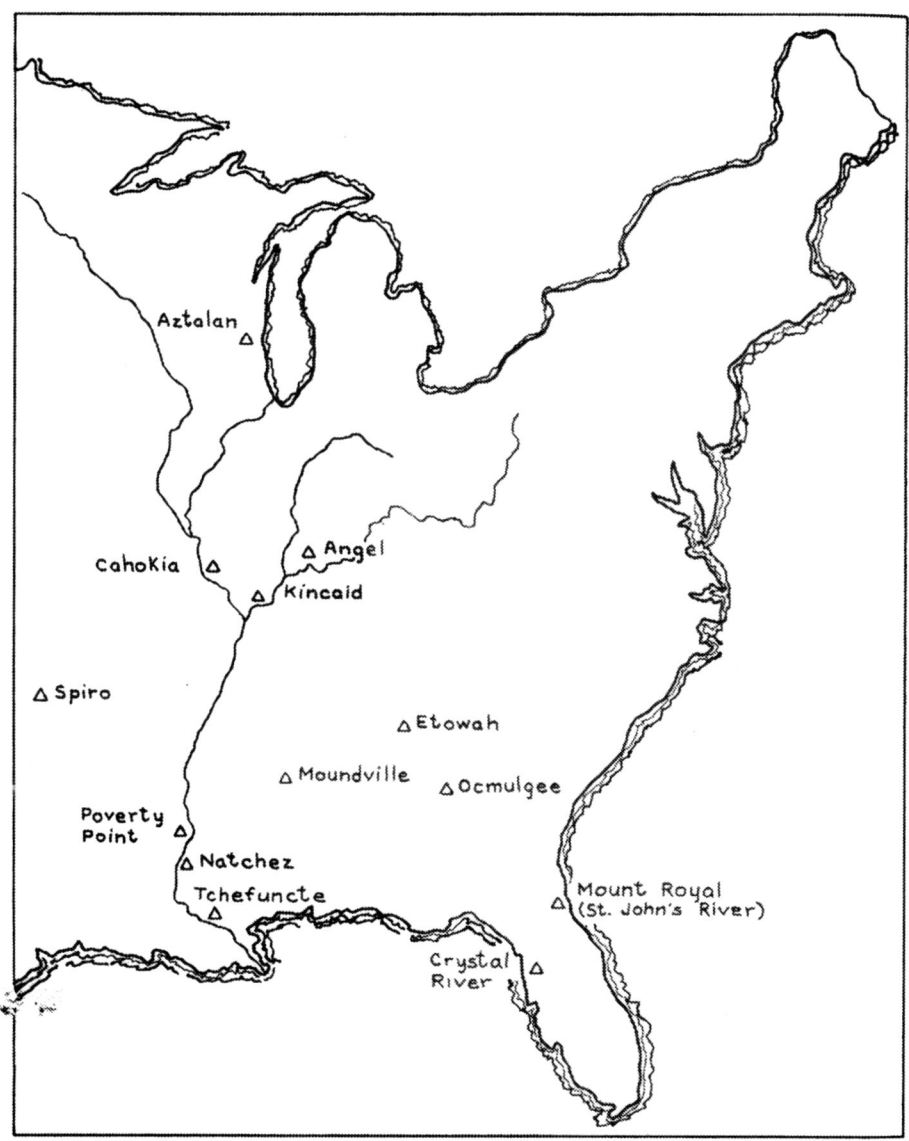

Temple Mounds Sites Map

Huge temple mounds dominated **Ocmulgee** on the Macon plateau in Georgia, a major Mississippian chiefdom between 900 and 1100. **Great Temple Mound** once supported three separate wooden structures. A steep stepped ramp climbed the front of this ceremonial pyramid. Nearby **Funeral Mound** encased seven levels of sacred Mississippian elite in carefully built log tombs. Perhaps most tantalizing is the huge subterranean earth lodge descending deep into the center of this site. Inside the cool recesses of this ceremonial chamber an ancient clay floor is surrounded by 47 individual seats in a clay bench. The whole room is about 42 feet in diameter radiating out from a sunken central firepit. An immense eagle effigy with a forked eye dominates this room. Along with this, six temple mounds and one burial mound complete this major site. Nearby **Cornfield Mound** covers the remains of a sacred cornfield that must have held certain ceremonial importance in the planting cycles. It was built up to eight feet high with native earth and covers rows of mounded dirt.

Craig Mound at **Spiro** has been reconstructed (after vandalism) and measures 300 by 115 feet and stands 35 feet tall. This temple mound was used for more than 800 years and held more than 1,000 bodies buried within it. The burials seemed to be very important ruling elite surrounded with amazing finely crafted artifacts. Spiro once controlled 200 to 300 ceremonial centers throughout the Arkansas River valley and was a hub of commerce for more than 100 years. This great center gradually declined in population and leadership importance, and was abandoned about A.D. 1450.

Angel Mounds in Evansville, Illinois, mark another Mississippian chiefdom that flourished between 900 and 1600, with a population of more than 3,000 people. This regional center covered over 103 acres of rich farmlands close to the Ohio River. An imposing stockade once protected this town site on three sides opening out to the river. **Mound A** towers 44 feet high above the town and covers four acres. Angelino elite raised their temples atop these manmade mountains, surrounded by perhaps 200 dwellings. As they gradually used up their natural resources, the Angelinos moved into other regions and may have joined other Mississippian settlements.

9
A Monumental Task~How Did They Build Those Mounds?

○ ○
We Indians think of the Earth and the whole Universe as a never-ending circle, and in this circle, man is just another animal. The buffalo and the coyote are our brothers; the birds, our cousins. We end our prayers with the words "all my relations"—and that includes everything that grows, crawls, runs, creeps, hops, and flies.

—*Jenny Leading Cloud, Rosebud Sioux*

> Mounds mark over 4000 years of American prehistory. The ancient mound builders created thousands of different mounds and complex earthworks in many shapes and sizes across the central and eastern United States. To do so, they moved millions of pounds of earth. The mound builders' technical achievements and engineering feats are still amazing to us today. All work thousands of years ago was fueled by human labor. How long did it take to build some of the biggest mounds?

The rituals of mound building flourished four millennia in America. The technology of ancient mound builders enabled them to develop the engineering skills to build enormous earthen mounds in many shapes. The ancient Indians placed priceless objects of art in the mounds along with the bodies of their deceased. Thousands of Indian mounds and complex earthworks were constructed across the central and eastern United States. Yet for all this diversity, the mounds usually take three major forms: conical for burials, platform for rituals, and effigy for honoring the earth. Presumably all mounds honored the earth amd celebrated the mound builders' worldviews and religious beliefs.

Building the Mounds

The mounds were made entirely of earth built up over continuous stages of construction. The mound builders obviously visualized each engineering feat in careful detail and the labor must have been intensive over the spring, summer, and autumn periods of good weather, before the earth froze in winter. Some mounds were built up over the course of generations and some took over 100 years to fully create.

First, people had to loosen and dig the soil using stone hoes hafted to long, sturdy hickory branches; perhaps they also used deer and elk antlers and hardwood digging sticks. Next, they may have used bison (buffalo) shoulder blades and elk scapulas hafted onto sturdy branches like shovels to load the fresh earth into burden baskets. Then teams of workers carried each load of earth, perhaps weighing 50 pounds or so, in large tightly woven baskets slung on their backs and fastened by burden straps across their foreheads or shoulders. Perhaps they used bamboo (cane) splint baskets, or baskets woven from oak, ash, maple, or basswood splints or willow rods. Perhaps four men at a time dragged heavy loads of earth piled on top of large buffalo robes, each one holding a leg, until they had transported many tons of soil up steep inclines to construct major mounds.

Carrying or burden baskets are still woven of river cane and wood splint materials by today's Indians in these regions. Tumplines, or burden straps, were readily finger woven from peeled milkweed and dogbane bark fibers and from the peeled inner bark of hemlock, ash, hickory, maple, and oak trees. Mound Builders' work tools and baskets do not turn up in archaeological sites, so we must imagine what they used to accomplish these amazing feats of human engineering.

Imagine what monumental tasks were involved in planning and constructing the mounds. A constant stream of workers would have been necessary to construct the many diverse mounds and earthworks which these sophisticated cultures have left. Perhaps the work grew slowly and naturally with the first sacred burials; then additions were made to each new phase of burial. People could see great monuments growing up on the land and the reverence for this way of honoring the dead grew with considerable strength.

Mound building could best be accomplished during the three seasons when the earth was not frozen, and during fair weather, not rainy wet conditions. Hundreds of workers were necessary to transport the countless tons of earth, while hundreds more were needed to dig it and many others to stamp it down firmly with their feet. This was hard labor, yet it must have been honored and distinctive work in mound builders' society. (Perhaps it was a bit akin to working on Wall Street in today's society.)

Perhaps whole villages turned out to work together to create their fine conical burial mounds, enormous platformed temple mounds, and striking effigy mounds, and other long, linear earthworks. Mound building must have been orchestrated with intense division of labor and every able-bodied individual, including children, might have had a role in some aspect of the construction. This would make good sense considering that the mounds were obviously enormous spiritual and ceremonial creations and central complexes within these dynamic societies.

So much soil was dug up and transported that this left huge depressions in some areas called barrow pits. The barrows soon filled with water and became bogs or ponds. Wild geese, ducks, swans, and other game birds came to nest in these areas; moose, bison, and deer came to eat the tasty plants and drink at these watering holes; frogs and salamanders would breed and live there; turtles, crawfish, and diverse fish populated these environments—then hunters would come to selectively take their seasonal catch. Like the quarries where rocks were harvested, these barrows were tremendous resources for the mound builders, who also harvested the wild plants from these areas for foods and medicines.

Inside the Mounds

The mystery of the mound interiors has long fascinated the people who study them. Each type had a different purpose and was constructed differently. Each seemed to have served distinct and separate ceremonial needs within mound builders' societies. Many hundreds of mounds have been investigated in order to try to learn their primary importance as well as how and why they were built. More is known about some mounds than others, and some continue to be mysteries.

The conical mounds used for burials contained one or more bodies, grave goods, and (sometimes) ashes from cremations, while the platform and effigy mounds did not usually contain burials or other offering objects, except in rare instances. Clearly the conical burial mound was a vital cultural practice for thousands of years among the ancestors of today's Native Americans. We glimpse inside some of these ubiquitous mounds and gain valuable insights into the sophisticated native societies who created them.

Some of the Most Extradinary Mounds

The earliest known mounds and earthworks were built during the late Archaic in Louisiana at **Poverty Point**. A series of mounds and effigy earthworks were constructed, beginning in about 1800 B.C., during a millennium of activity that look like a large bird with outstretched wings hovering near the Arkansas River. This represents one of America's earliest and largest prehistoric earthworks. No human burials were made there, but an important human cremation was part of the early formation of the central mound. Millions of cubic feet of earth were moved to construct the concentric circles of the bird's wings stretching out two-thirds of a mile. Today this site is protected as a 400-acre archaeological park. More than 100 similar sites related to Poverty Point have been recognized in Louisiana, Mississippi, and Arkansas, and reach as far east as Florida.

Plains Indians Burial Mounds

About 1000 B.C. early ancestors of the Arikara, Hidatsa (Minatare's), and Mandan began to settle in communities where they made pottery, began early forms of farming, and constructed burial mounds. While their culture and ceremonialism was less elaborate than the eastern mound builders, these Plains Indians developed similar traits and lifestyles. They developed more settled lives and built permanent earth lodges by about A.D. 1000 at several sites along the Knife and Missouri Rivers.

Villagers continued to hunt the buffalo, yet they relied increasingly on farming an early-maturing corn, along with squash, pumpkins, beans, tobacco, and sunflowers planted in the fertile floodplains. These villages were also in fine positions to be major trading partners for the lucrative exchanges made between far-flung tribal communities for over a thousand years.

Knife River Indian Villages National Historic Site in Stanton, ND exhibits the remains of three Hidatsa Indian villages, along with other cache pits and lodge depressions. Lewis and Clark stopped at this site in October 1804 when Sacajawea joined their expedition. They wrote extensively about these generous people. Today the Arikara, Mandan, and Hidatsa are the Three Affiliated Tribes living on the Fort Berthold Reservation. Little remains of the prehistoric burial mounds.

In old times we Hidatsas never made our gardens on the untimbered prairie land, because the soil there is too hard and dry. In the bottomlands by the Missouri, the soil is soft and easy to work. My mothers and my two grandmothers worked at clearing our family's garden. It lay east of the village at a place where many other families were clearing fields.~Maxidiwiac, (Buffalo Bird Woman) Hidatsa, was born about 1839 in an earth lodge along the Knife River, North Dakota.

Double Ditch, the largest prehistoric Mandan village site in North Dakota may have supported and enclosed more than 3,000 people with deep ditches and earthen mounds. It was probably abandoned after the small pox epidemic of 1781 as Lewis and Clarke found it empty when they visited in 1804.

Amazing Technology

The Native people who built the mounds had developed an amazing degree of technology. Some became highly skilled at making fine pottery, while others were talented at making fiber and cordage, and still others at weaving plant fibers into burden straps (tumplines), fishnets, cloaks, bags, and mats. Many fiber plants were also powerful medicines, and some fiber makers were also medicine men and healers.

Some mound builders were especially skilled at shaping and drilling stone and shell beads. Mound builders had a distinctive sense of adornment. Their fashions may have been light on clothing but remarkable on jewelry: earrings, pendants (gorgets and breast plates), necklaces, feathered hair ornaments, and beaded medicine bags. We can only imagine how many other objects of adornment they may have developed.

Mound builders made beads from many different objects found in nature including fish vertebrae, pieces of shells, soapstone, and semiprecious stones. Yet their greatest passion was for freshwater pearls! Countless thousands of freshwater pearls have been found in burial sites, as well as pendants made from the pearly, iridescent inner shells of mussels, which produced the pearls. Some pearls were set into copper and bone and stone objects; some pearls were burned. It is unclear what special meanings pearls had for these talented ancient mound builders.

Quarries and Bogs

Flint, chert, pipestone, and soapstone were useful, desirable stones sought by Native American craftsmen. Flint and chert are dense, hard stones suitable for spearpoints, arrowheads, and fine cutting tools. Pipestone and soapstone are soft, easily shaped stones desirable for carving pipes, beads, bowls, and other useful items. Cahokians quarried a type of glassy chert from nearby veins in the earth.

There were also nearby salt mines where the mound builders obtained salt in abundance to use for trade with other groups of people.

Bogs are unique wet places, like marshes beside a river or lake. But bogs are usually inland "bowls" where water is trapped over time. Special animals and plants live in bogs, especially certain medicinal plants, like cardinal flower to treat heart problems, blue flag as a stomach tonic, bulrushes for food, basketry and mats, and bog milkweeds for kidney and heart medicines. In the woods at the edge of bogs could be found the pennyroyal and pipsissewa vital as stomach medicines, and used to reduce fevers and treat snakebite. Both bogs and quarries were unique ecosystems where the mound builders made repeated visits and often camped and harvested important resources.

Cardinal Flower, *Lobelia cardinalis*--valuable perennial medicine plant, especially for heart problems & 'love medicines'

They also harvested herbs and medicinal roots and bark from bogs, marshes, and wetlands for additional health needs. They could frequently gather enough extra to use as desirable trade items. These wet regions produced willows for pain relief (and basketry), boneset for colds and to strengthen the immune system, sweet flag to aid digestion and treat sore throats, pitcher plants to relieve colds and flu, and swamp milkweed to treat heart problems, stroke, and high blood pressure. Some of the native orchids and alders that grew in the bogs were valuable to treat everything from snakebite to respiratory distress and asthma, and also useful in birth control formulas, which mound builders had knowledge about.

Pipsissewa, *Chimaphila umbellata*–small perennial (evergreen) medicinal plant used for many therapeutic needs.

Lithic technologists were and are special craftsmen who work stone. These exact techniques had to be developed over time and included countless hours of practice. But first the right stone had to be located.

Native toolmakers knew the best quarries and journeyed great distances to get flint and chert for special tools, weapons, and ceremonial objects. Some of these rock quarries had their own special energy from so many people coming to harvest from them over thousands of years. Flint Ridge quarry, just a few miles from the Newark Mounds in Ohio was famous for its Ohio pipestone.

At Poverty Point the ancient mound builders created tiny owl effigies out of jasper and polished them. Finely worked arrowheads and spearpoints were hammered (or pecked) out of flint by ancient hunters. Toolmakers hammered stone or antler against stone to make obsidian blades, granite hammer stones, adzes, and axes, and countless other stone items for their needs. They fashioned needle-thin flint drills to pierce shells and pearls, which were used for adornment.

The stones are earth's bones and alive with beauty. Some stones bring up the essence from deep with Mother Earth's womb, and they are more valuable, like the flint, chert, and garnets. Some storytellers say that garnets were the Original Little People, made by the Great Mystery before the beginning of time. Garnets are a deep blood red and have eight perfect, flat, shiny sides.

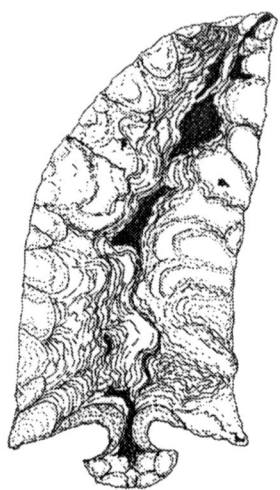
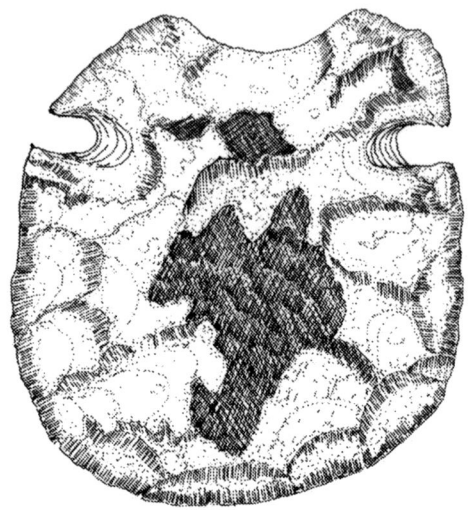

Hopewell Obsidian Ceremonial Blade & Cahokia Chert Notched Hoe Blade

Hopewell Mound Group, Ohio

[lashed to a wooden stick with leather cord & animal glue] Cahokia Mounds

The Stone People—An Ancient Algonquian Story

This early Algonquian story from the Eastern Woodlands is especially important in a study of "Stone Age people"—whose technologies still dazzle us with their fineness. The "Stone Reference Guide" sidebar takes us deeper into our awareness for "Mother Earth." The ancient mound builders knew the stones and rocks of their environments and traded for exotic stone from distant locations. Many Woodland Indians shared stories about an early race of "stone people"—a mythical race of beings created before real people.

Back in the time when the world was new, the Great Mystery was busy shaping the mountain ranges to be the backbone and belt for Mother Earth. Much of this material came from deep within her to beautify her surface and make the world radiant. Great

Mystery enjoyed working and shaping each thing, putting knowledge and love into it all. This is how it came to pass that the Stone People were created, long before other living things. Great Mystery wanted someone to hold the knowledge of these deep origins during the rest of Creation Time. The Stone People volunteered. Many other rocks also volunteered.

Great Mystery shaped the Stone People in most agreeable deep blood-red molten rock. Flecks of other precious stones were mixed into their bodies and seasoned with the dust of ancient earth knowledge. When everything was just right, Great Mystery pressed the Stone People through amazing metamorphic treatments, which melted them and then birthed them each with perfect octagonal shapes. They had eight perfect, flat, shiny sides around their whole—almost round—forms.

Some of the countless Stone People popped out whole and rolled around in the new soil of Mother Earth. Others chose to rest in great clans and villages with their relatives in sparkling schist and granite gneiss rocks.

Millions of years passed and Great Mystery decided to make bigger people of flesh and bone, who would admire the Stone People from long, long ago. Our ancestors discovered the Stone People and collected them for their medicine bags and healing rites. Many people wore them as beautiful adornments. Children gathered Stone People to use in their games.

We call the Stone People garnets today, and honor them as semiprecious gemstones. Garnet is the birthstone for Aquarius, those people who are born in late January and early February.

Mother Earth still wears much of the granite schist in New England as part of her ancient backbone, which is the mountain belt that we call the Adirondacks and Appalachians. Garnets are found, usually in pockets and veins, around the world.

Modern geologists tell us that garnets hold the knowledge of their ancient path through time as they come from the core of the earth from millions of years ago. Today we know they represent the late Silurian or early Devonian Age, estimated to be about 350 to 400 million years ago. We continue to consult these beautiful Stone People to learn what else they might be willing to tell us. Some of us even wear them and carry them for good luck. [Gratitude to Dr. Chris A. White, Geologist, UMASS-Amherst.

Weapon Makers and Tool Makers (Lithic Technologists)

Men skilled in constructing stone tools and weapons were in great demand in early native societies. Hammering and pecking stone against stone, and antler or bone against stone, required technical skills along with knowledge of different rocks and how they would fracture. The hardest kinds of stone, like flint, chert, and obsidian, were most sought for tools and weapons. Quartz was also used and quite common, but it fractures in odd ways and was not as dependable for the toolmakers. Quartz commonly fractured into sharp spearpoints and arrowheads.

Hard river cobbles were laboriously shaped into hammerstones, adzes, gouges, and axes for the heavier work needs like felling trees and hollowing out logs to make dugouts and large feast bowls. Talented technologists easily made stone, bone, and antler tools. Deer antlers and the hip bones of deer and elk were fashioned into hoes, attached to long sticks, as handles, in order to dig the gardens for planting and scoop up the earth to fill the burden baskets of the laborers who worked to build the mounds.

A STONE REFERENCE GUIDE:

Chert: a hard rock of crystalline varieties of silica.
Feldspar: very hard, fine-grained crystalline rock abundantly found.
Flint: very hard, fine-grained quartz that sparks when struck with steel.
Garnet: a semiprecious gemstone of crystallized silicate minerals. Although usually dark red in color, it can be brown, black, green, yellow, or white.
Gneiss: a banded type of granite whose minerals are arranged in layers.
Granite: common, coarse-grained igneous rock made mainly of quartz and mica.
Mica: a common silicate mineral usually found in igneous and metamorphic rocks; it readily splits into flexible sheets with shiny, mirrorlike flat surfaces.
Obsidian: lustrous volcanic glass, usually black or banded, displaying curved, shiny surfaces when fractured. It is extremely sharp!
Quartz: the hard crystalline mineral silicon dioxide; and in pure crystals such as agate, chalcedony, chert, flint, opal, and rock crystal.
Schist: various coarse types of metamorphic rock in layers that can be flaky.

Metamorphic means changed by great heat or pressure. Silicates are common compounds of silicon and oxygen, and other materials that occur in most rocks except limestone and dolomite. This forms the basis of common glass. The various rock quarries must have been as valuable to the Indians as the coal mines and oil rigs are to modern folks needs.

Amazing changes in the continent's heartland continued to take place as Native People prospered and evolved. Natural resources provided abundant opportunities for native cultures to flourish. Dynamic cultures living close to nature, with profound reverence for their leaders and ancestors, constructed mammoth earthworks as living monuments on the land. These magnificent mounds certainly imbued the earth with powerful energies from the ancestors' spirits. As populations increased, many native groups settled in growing cities with numerous satellite villages surrounding central trading and ceremonial centers. Rivers continued to be the main arteries of life and the mound builders were great navigators and traders.

10

Cahokia~City of the Sun

∘ ∘

What is life? It is the flash of a firefly in the night. It is the breath of a buffalo in the wintertime. It is the little shadow, which runs across the grass and loses itself in the sunset.

—Crowfoot, Blackfoot warrior, 1890; his last words

> The architecture and physical layout of Cahokia, and the social and political structure within the Mississippian culture are examined. Who ruled and how? Artifacts, tombs, and reconstructions show us details about these mysterious ancient societies and their leaders and economy. Cahokia's agriculture, trade, and manufacturing were central to this city's prosperity. Discover why Cahokia became the largest pre-Columbian city in America. Scientists named Cahokia for a tribe of Illini Indians who camped and lived near this site, and might have been distant descendants of the Cahokians. We do not know what these ancient people called their city.

A magnificent prehistoric city evolved where earlier tribes had also flourished. Cahokia was built on the broad flat floodplains where the Mississippi and Missouri rivers come together in America's ancient heartland. Native People had been camping and living here for thousands of years, long before agriculture became a staple in prehistoric life. The early Woodland Indians hunted, fished, and gathered the seasonal bounty enjoying semi-permanent settlements across this region. The fertile floodplains opened out into rolling meadows leading to dense woods, where hunting, fishing, farming, and mound building flourished. Cahokia Creek meandered through fine small settlements of bark-covered wigwams.

Late Woodland Indian villages were well established in this area before A.D. 700 These mound builders fished and hunted throughout the region, gathered wild plant foods and began cultivating early types of corn and squash. They enjoyed the wealth of surrounding resources and built small compact villages in this place. As these earlier cultures prospered, they set the stage for a grander culture with a bold new economy that was sweeping the river systems and fertile bottomlands. The ideas for a great society may have begun in the east 200 years earlier. Yet opportunities at Cahokia would allow them to surpass everyone else. This evolved into the heart of Temple Mound Building societies embracing bigger ideas for farming and celebrating life and death. The social order centered around the ruling elites and people flourished at every level of life and were honored in death.

Cahokia became the vital hub of an enormous, lucrative trading system that was its life force for growth. Abundant wildlife and fertile soil made life easy here and its desirable location on major trading routes marked this pre-Columbian city for great success. Cahokia became the "Crown Jewel" of the Mississippian Temple Mound building world. From about 900 to 1250, Cahokia ranked as the largest, most powerful city north of Mexico. There were a number of remarkable assets surrounding this location, including the conjunction of various ecological zones allowing for plentiful resources to enrich Cahokian life. The river systems facilitated both north—south travel, and east—west exchanges, insuring that Cahokia would be a major center of trading, commerce and political life.

This was an ideal location in the middle of the American Bottom, one of the finest agricultural sites in the country. The "bottoms" are low floodplain deposits along rivers, where deep sandy loams accumulate. These choice soils are very light and fertile, and easily worked with hoes and digging sticks for farming. The American Bottom was a gift of the Missouri River from its long journey across the plains from Montana, where it picked up quantities of topsoil and silt. *Missouri* is an Indian word for "muddy water" and this blessing created some 125 square miles of bottomland.

> Most of the world's great ancient civilizations arose along rivers for ease of transportation and rich water harvests, and where the most fertile flood plain soils were available for successful farming. These loose alluvial soils were also excellent for mound building. (The city of St. Louis, Missouri, grew up just eight miles west of Cahokia across the Mississippi.) Mound building was not only sacred glorification of revered dead, but monuments to greatness in Cahokia.

By around 850 a cosmopolitan culture had developed. The people had bigger ideas and developed better technologies than their predecessors. Cahokian women tilled larger fields of corn, squash, pumpkins, beans, tobacco, and sunflowers with flint and bone hoes. They were able to grow a stronger, more productive type of maize (corn) imported from Mexico. In some sheltered areas they could plant and harvest two crops of corn each year. They could grow more food and this helped develop a bigger, wealthier economy. Cahokians traded for new strains of corn, squash, and beans from their southern neighbors. As farming increased, they established larger, more sophisticated settlements wherein people could develop specialties like pottery making, tool, fiber, and jewelry making, etching of shells, and pearl harvesting and ornamentation, and these products enhanced Cahokia's trading base.

Hunters using bow and arrows, snares, and traps bagged bison, deer, wild turkey, geese and ducks. Fishermen worked the rivers and lakes for the multitude of fish, eels, mussels, and turtles seasonally available. Women and children harvested the seasonal bounty of wild berries, seeds, nuts, herbs, and medicinal roots. Dugouts and canoes packed with Cahokia's finest furs and produce left for distant trading centers and returned laden with exotic raw materials to continually stimulate this economy.

Salt trade became an important commodity among Mississippian farmers by about 900. Hunters and gatherers rarely needed salt in their diets, which generally contained sufficient minerals. Yet farming cultures grew to need salt, not only for taste, but to overcome a potentially fatal salt deficiency that could develop. Most salt was made from the waters of saline springs located in four primary areas: Cahokia's region; and near the Wabash River between Ohio and Indiana; near the Red River in Louisiana, Arkansas, and Texas; and along the mid-Ohio River and its southern tributaries. Native people in these regions were the salt makers and others had to trade with them for this ancient flavoring, preservative, and mineral additive. Cahokia was a major salt producer.

Cahokia, City of the Sun

Imagine the heyday of Cahokia in the year 1150. This vital cosmopolitan center is strategically located to grow bigger than any other Temple Mound Center and everyday life here seems to sparkle with a great range of activities. Young children run and play together amidst clusters of women cooking around open campfires, laughing and talking together. Many other women are out working in their cornfields, hoeing and tending sizable gardens. Babies and toddlers stay near their mothers. Men work in separate, individual locations pecking stone against stone to create work tools and weapon blades. Small slivers of rock fly off from their pecking work and litter the sur-

rounding area. Other men sit in a cluster on the ground shaping stone pipes and talking in low voices.

In the distance a group of potters work freshly gathered green clay, sifting, cleaning, and kneading it. Others nearby are shaping fine cooking and ceremonial pots, cups, pipes, and figures in clay. A little distance away from them another group of potters are pit-firing clayworks they created a week ago. They work near a deep recess in the earth lined with clay and firewood. This will take all day.

Crossing a distant plaza, teams of hunters return with freshly killed deer and wild turkey, while young men are spear fishing in Cahokia Creek for catfish, sunfish, and bass. They and their women will skin and prepare the meats for distribution. Fine turkey skins, like the best deer hides, will be tanned with the feathers on them. They will be fashioned into long turkey feather capes, which are excellent rain gear. These capes are generally only worn by leading men: sachems, shamans, and sub-chiefs.

A small group of men and women work together constructing a new house.

Four large corner posts have been securely planted in the ground about 10 feet apart. This will be a modest, square house, yet some in the area are larger rectangular homes. Smaller poles are set in between the larger corner posts, and big roof poles are placed and secured with lashings overhead. Nimble young boys will climb the wooden ladders and stand on their father's shoulders to lash bundles of rushes and cattails to the cross poles, making a fine roof using abundant renewable resources. Three young women are busily weaving supple alder, willow, and hazel rods in between the side poles, creating the house walls. This is like weaving a huge standing basket. Their brothers follow along behind them, two on the inside and one on the outside daubing mud and clay over the woven branches to make smooth, thick walls. There is always an abundance of mud here. This dries hard like bricks. [Centuries later the Spanish would call this adobe architecture when they saw similar types of houses in the Desert Southwest.]

The majority of people in this and surrounding villages are working to transport more earth onto the top of a nearby temple mound. Teams of carriers bend under the weight of 60-pound pack baskets filled with earth on their backs. They climb up the steep sides of the mound to deposit their loads on the top. Mound dancers are busily spreading and dancing on the new soil with their bare feet. They use deer antlers as rakes to smooth the soil evenly over the flattened mound top. The mound builders are digging soil from the edges of Loon Pond, which is low because of dry conditions. It is a long walk to the temple mound, but hundreds of carriers make this trek many times daily. They are young and physically fit; they enjoy watching this new ceremonial mound grow. Everyone is barefoot and wears relatively little more than soft deerskin skirts and vests. It is the only way to go about except in the coldest winter times.

By 1150 Cahokia covered about six square miles and was home to perhaps 20,000 or more people. A series of open plazas, groups of houses, and small gardens subdivided this magnificent city. Between clusters of small post and daub houses there was a parklike landscape. Blueberry, wild plum, currant, gooseberry, and huckleberry bushes grew well here along with wild strawberries and diverse wild medicinal plants. Wild grapevines climbed low trees and produced abundantly in late summer.

Various temple mounds stood at a distance from one another and some supported large longhouses, or temples, atop the broad flat surfaces. The long temples were built of post and daub architecture beneath high crowned cattail and rush roofs. These towered over the smaller post and daub cottages of elite Cahokians clustered across the flatland below. The temple mound dwellers looked down on life below them and saw agreeable landscapes. Footpaths filed neatly between lush patches of wild grasses and sweet herbs sending up fragrances in hot sun. At the center of the city facing south stood gigantic Monk's Mound, which was connected by a Grand Plaza to Twin Mounds. Cahokia Creek meandered just beyond the back of the Great Temple Mound and marked the back boundary of the great palisade with its lush marshes and small islands. Each mound at Cahokia had a special name and several individual clans of people devoted to its upkeep.

Monk's Mound, the **Great Temple Mound**, was the center of life in Cahokia. Its 16-acre base supported its 100-foot center height making it the world's largest earthwork. Manmade by thousands of laborers working without draft animals or wheels, Great Temple Mound was probably built by basketfuls of earth in perhaps 14 stages between 900 and 1100, and contains 22 million cubic feet of earth. The summit supported a huge wooden building, perhaps 50 feet high, measuring 48 feet wide by 105 feet long. Several additional fine buildings stood juxtaposed to the great temple, and this was all surrounded with a low palisade wall opening out to the grand center stairway leading down from the terraced summit to the lower terrace, and on down to the plain below.

Wild Strawberry, *Fragaria virginiana*–fruits for foods & leaves for teas & medicines

The elite ruler, The Great Sun, and his guardian sun priests lived there and governed Cahokia's ceremonial and practical world from this elevated position. They set the planting and harvesting times; they determined when each fertility rite and harvest festival would occur, and when sacrifices needed to be made. In 1150 The Great Mound was a beehive of activity. The Great Sun and his aristocratic family and priests ruled over this imperial mecca. This, like all hereditary lineages at Cahokia, passed through the female line. Women held the power and decided who would lead in each clan. Cahokia was the City of the Sun and sun symbols were everywhere and on everything.

A great **Sun Calendar**, considered the American Woodhenge, was first constructed at Cahokia in 1100. It stood about 1,000 yards west of Cahokia's Great Temple Mound. The Temple Mound astronomers could read the Sky World and they plotted sacred seed planting times and ceremonial honoring feasts by observing the sun's alignment within this huge earth clock. A 30-foot tall cedar center post was painted red and sunk into a deep hole, and tamped in securely at the center. Another 48 telephone-sized posts were spaced out in a large perfect circle around it, and also sunk into deep holes, and tamped down securely. These cedar posts were also painted red and stood perhaps 15 to 20 feet tall. Four key posts were exactly aligned with the four cardinal directions of north, south, east, and west. The remaining posts were evenly spaced in between them forming a circle some 410 feet in diameter. Here the calendar keepers could predict the coming solstices and equinoxes, and other important dates. This must have looked spectacular from atop the Great Temple Mound.

The great **Sun Calendar** was accidentally discovered in the early 1960s as a team of archaeologists worked ahead of bulldozers putting an interstate highway through the heart of Cahokia. They called this great triumph of American Indian science and engineering The American Woodhenge after Stonehenge in Great Britain. Additional "sun circles" were eventually found in the outlying regions also beyond the palisade. The largest might have once sported 60 cedar posts.

Radiocarbon dating indicates that the woodhenges at Cahokia were used for perhaps 100 years, between A.D. 1100 and 1200. Today the function and use of these amazing structures seems to have been lost. Scientists who have been studying them believe that the sun circles might have been similar to the Plains Indians "world center shrines" sym-

> bolizing a certain cosmic order, and perhaps used for vital annual reckoning of key events and fertility rites.

More than a hundred other mounds stood outside the great palisade wall. Trading rivalries and warfare appear to have been a growing factor in Cahokian life, and across the Temple Mound Builders world. The ceremonial heart and center of Cahokia was fortified against intrusion, which set it apart even more from the many people whom the leaders served. Only the Cahokia elite maintained houses closely ranked together within the great wall. All of the large gardens and cornfields also lay outside the palisade.

This 15-foot tall stockade or palisade wall surrounded the city by 1100. The palisade walls were interwoven with saplings of willows and dogwood branches tightly packed and covered with river clay, like most Cahokian homes. This two-mile-long encircling palisade was made of 15,000 oak and hickory logs. Cutting so many trees must have made quite an environmental impact in a region dependent upon firewood and building materials.

The broad bottomland was dotted with giant oaks, maples, and clusters of sassafras and ash trees. Numerous conical burial mounds rose here and there. Grouped around these various features were the small houses where average Cahokians lived. Their small post and daub bungalows reflected, to some extent, the housing of elite Cahokia. Their broad farm fields spread out for miles well beyond the city on the surrounding floodplains, interspersed by ponds and lakes that formed in the enormous barrow pits dug by Cahokian laborers, who carried the soil in burden baskets on their back to build the massive mounds. This work continued periodically.

> *Mineola*, a place name, comes from the Algonquian word for "palisaded village."

~~~~~~~~ Step back in time ~~~~~~~~

Imagine a mild spring day in Cahokia in 1150: The Great Sun warms before an early altar fire in the earthen center of his ceremonial lodge atop The Great Temple Mound. He has returned from the clay courtyard just outside his door, where he sings the sun up each day with prayers and special appeals for his people's success. He lingered this morning looking at the tendrils of smoke rising from many small cook fires below. He smelled and tasted the fragrances of the day and watched the sky for signs of rain. It was clear and bright.

Inside he savors the fragrances of his own lodge. He watches his wives and children do their morning chores. The firelight flickers and shines upon the mica sheet windows near his head. These are set into the lodge walls at intervals of about four feet. The ceremonial lodge is long and narrow with a high bowed roof. Its creamy colored clay walls are covered in many places with finely woven cattail mats of beautiful colors and geometric patterns made by his wives. Large deerskins and buffalo skins hang on the walls too, carefully painted with scenes of the past few years. Each one is like a calendar account of the major events and trading excursions experienced at Cahokia during 13 moons. The Great Sun's scribes and shamans paint these robes to mark key events and keep vital records. He is surrounded by beautiful gifts from far-flung trading partners and lesser Suns across the country. Exotic seashells are placed around the lodge, and fine wooden carvings hang from the wooden posts by leather thongs. His great stone pipes rest along the rim of his smoking altar.

The Great Sun and his many Sun Priests are concerned about the great need for rain. Spring is late and cool, and there has been no rain in many weeks. The moon of planting time is here and the earth is parched and cracked. Everything depends upon rain for its abundance! Drastic measures must be taken.

The Great Sun has called for a day of drumming, dancing, prayers, and feasting. Everyone must participate and all are willing. Hunters have brought in extra game~migrating geese, ducks, and swans. His favorite wife is making him a new head crest of white swan feathers. Women have been preparing foods and cooking for the past two days. Special offerings of foods and animal blood are made to the Spirits of Rain over prayers of intense appeal.

Cornmeal blessings are scattered over the dance ground of the great plaza below by the sachem. The Great Sun calls for the dancers to begin and the drummers who encircle them begin pound out the heartbeat rhythm. The BOOM~BOOM~BOOM of these ceremonial drums calls the last stragglers together below. Many people watch in clusters as the ceremonial parties begin their magic dressed in their glorious clan attire.

Lines of the best dancers in their finest regalia stretch out across the plaza in their classic swaying dance, gently shifting from one foot to the other, as they pick up each foot and place it back down in rhythm with the drumbeats. Their bare feet are massaging Mother Earth and their dances are their prayers for rain and fertility and another good growing season. The women wear fine, soft deerskin tunics over short deerskin skirts fringed and beaded with seashells and freshwater pearls. They wear strands of pearls around their necks. Bright feathers are braided into their long hair. They hold gourd rattles in their right hand and an ear of dried corn in the left hand.

The men wear short, fringed deerskin kilts and finely woven sashes across their bare chests. Bright feathers are braided into their long hair, along with sweetgrass. They hold gourd rattles in their right hands and spruce branches in their left hands. They wear strands of pearls and bear claws around their necks. Several young girls dance into the plaza attired as butterfly maidens, come to pollinate the spring crops and flowers. They wear wreaths of flowers in their hair and are dusted with yellow pollen and cornmeal. Several young men dance into the plaza along with them as young antelope, deer, and buffalo dancers~fresh for the new season. They wear headdresses with the young horns of these animals. Each dancer chants along with the drums. These beautiful young dancers prance to beseech the natural world to again bring abundance to their people.

Dancing, drumming, and chanting continue with a steady, easy rhythm all through the day, almost without pause. As one group of fine dancers tire (after about an hour) they walk back to their clan lodges for rest and food, while another group of dancers takes their place. There are so many who wish to participate in "teams" this way. There are hundreds of dancers in the plaza at any given time. There are also teams of drummers who relieve each other so that the ceremonies continue like an unbroken thread. There is a mesmerizing feeling of high energy and power here!

By afternoon storm clouds are building across the western sky and lightning flashes in the distance. It seems as if Nature and Mother Earth respond with enthusiasm to these ceremonies. The drumming, chanting, and dancing continue with real passion!

The drummers gather under a long brush arbor shelter so they can continue drumming. Soft rain begins to fall on the ceremonies, which become a celebration! The chanting grows louder. The rain intensifies. The Great Sun and the Sun Priests look down on this with great approval. They are relieved that the prayers are being answered.

Soon the dancers are dancing in mud up to their ankles, their bare feet massaging the earth's changing moods. Mud is beautiful. Afternoon lengthens and the last rays of filtered sun catch a double rainbow standing over Great Temple Mound. This is their ultimate reward! The ceremony ends and everyone makes their way slowly home in pouring rain. They feel cleansed. The rain continues all through the night. Planting time will begin in two days after the Sun Priests give more cornmeal prayers to Mother Earth and Father Sky.

~~~~~~~~~~~~~~~~~~~~~~~

Cahokia's Satellites and Trading Partners

Cahokia was the center of an immense trading system reaching out in all directions. This complex chiefdom and capital city controlled commerce, art, politics, religion, and healing throughout a vast region. The network of exchange was vital in Cahokia's life. It was a powerful engine driving the economy. In the 1100s the city's population rivaled that of the distant city of London, England, and the Temple Mound Builders had far-reaching influence.

A number of satellite communities stretched along the American Bottom, which was densely populated. Archaeological evidence gives us some special insights into some sites no longer apparent. The Mitchell site lay at the northern end of the valley near the mouth of the Missouri. Two other major sites stood where St. Louis and East St. Louis are today. Another neighboring site was Pulcher, where the valley narrowed. Perhaps four more mid-sized centers and 50 small farming villages were strung out across the valley. Each of these was within a 25-mile radius of Cahokia by 1000.

One satellite city of Cahokia was Aztalan, located in what is now Wisconsin. The Mississippians who lived here developed a thriving society beside the great Mississippi River. This huge complex was the hub of many unique Indian societies located throughout a broad region.

Satellite subregions reached as far east as North Carolina, Georgia, and Florida. These centers of ceremonial Mississippian cultures had developed their own specialties based upon their environments and locations. These were distant trading partners of Cahokia.

> An early map drawn by a member of the de Soto expedition (1539-42) shows a great region of the south dotted with palisaded towns. The Spaniards were impressed by the densely settled villages and abundance of food.

Men and Women Who Rule

Cahokia was ruled by great leaders who surrounded themselves with a group of special people. These were the social elite who helped them govern the growing populations. Political problems often developed, as in any great city, and Cahokia was not different. Strong leaders were needed to resolve differences over the building of roads, houses, trading centers, woodhenges, and more mounds.

Social leaders and subchiefs worked closely with the people and carried special messages from the ruler to the gatherings of different tribal centers. The paramount ruler, The Great Sun, presided over all major ceremonies of harvest and renewal, and at honoring times when the people would be asked to gather around the great temple mound. The Great Sun was considered so elevated that he had to be kept pure and above the ground. When he descended from the Great Temple Mound he was carried on a litter, a platform stretched between two long cedar poles, held by teams of chosen warriors.

Archaeological work uncovered the remains of cedar litter poles in an elite burial at Cahokia. Also the Spanish and French explorers' accounts confirm these patterns in later encounters. Both record early visits with Native American rulers known as Great Suns. Evidently they were surrounded in life and in death with luxuries. They were dressed in knee-length fur cloaks. The Spanish judged one marten skin cloak worn by a ruler in South Carolina to be worth 2,000 Spanish ducats, about $4,500. The Great Sun lived in a place called Palisema. His personal residence was covered floor to ceiling with fine tanned deerskins painted with many colorful designs.

> One Cahokian burial revealed the bodies of six individuals laid to rest amidst chunky stones, a large cache of arrowheads, sheets of mica, and large pieces of rolled sheet copper. One man was buried upon a blanket of marine shell beads. The cedar posts of a litter were also resting nearby.
>
> Chunky stones are about the size of a flattened grapefruit with convex sides. Since Mississippian times they have been fashioned to play the game chunky. Players roll a chunky stone along the ground and then throw a spear to the point where they believe the stone will stop rolling. This is still a popular game among the Creek Indians in Oklahoma.

Women were equally strong and powerful leaders among the mound builders, as among most Native American tribal groups. Native women were very far ahead of their European counterparts in this respect. They owned the homes and planting fields, had their say in electing the tribal leaders, and family lines descended through the female clan, not the father's male clan. For instance, when children were born to a turtle clan mother married to her badger clan husband, the children would all be turtle clan. When they grew up, they knew that they could never marry any of their turtle clan or badger clan relatives. They would have to select a marriage partner, from one of the other tribal clans. These concepts were fairly universal among most societies in order to prevent inbreeding and birth defects. This was quite possibly the case in Cahokia. It was recorded by the French in the 1700s that royal succession of the Great Sun followed a matrilineal (female) descent system. When a Great Sun died, a nephew, one of his sister's sons, succeeded him.

> **The Great Sun's supremacy** was explained in an old Natchez legend that told how man and woman descended to earth from the Sky World. Their role was to govern humans and teach them how to live better lives and practice rituals and ceremonies. The man (some say woman) was a child of the Sun. People were ordered to build a sacred temple raised up on a high mound. They created a great firepit within the temple where a piece of the Sun was brought to earth to give the warmth of fire. When these rulers died they turned into stone so that the earth would not corrupt them. Impressive stone figures of kneeling men and women have been found at Cahokia, and at sites in Georgia and Tennessee.

The Great Sun was exalted and chosen to live in the most elevated situation. He was believed to communicate directly with the sun, moon, and stars, therefore he had to live close to the heavens. Climbing up the great temple mound must have seemed like ascending the steps to heaven for the many Cahokians who had the privilege. When the Great Sun came down from his temple mound he was carried everywhere on a litter elevated on the shoulders of his most trusted warriors. It was somewhat like a "land barge" made of cedar poles trimmed and fastened together between two long cedar carrying poles.

Cahokia's social ranks probably supported a class of nobles allied with the Great Sun, perhaps they were Sun Clan People. Then there were Honored People, priests and warriors who served the elite from other important ruling clans. Perhaps the common folks were called "Stinkards," who came from other clans and outlying areas. Native people often kept captives, who were prisoners of war. Sometimes they were kept as slaves and others were welcomed into families as equals after a time.

People in early mound builders' groups who were born with birth defects or physical and mental challenges were usually treated with respect and considered to be especially "gifted." People with abnormalities were considered to have been "touched by the Creator" with special abilities beyond the obvious and normal range. Such beliefs persisted in many later American Indian societies—and still do today. Perhaps at Cahokia these special people became shamans and healers.

Drumming and chanting always accompanied every gathering. Huge cottonwood drums and cedar-rimmed drums, covered with elk or deer hide were played, like thunder, by the best drummers. A circle of other drummers playing hand drums would surround them, or stretch out in a long line behind the dancers. Everyone would chant and sing the honoring songs. Gourd rattles filled with hard kernels of popcorn created the sound of falling rain. These instruments were considered sacred and were more than simply musical devices; they echoed natural sounds and hoped to call in the blessings of sunny weather as well as abundant storms and rainfall.

Cahokian Artisans and Artists

Cahokia had many fine artisans—potters, basket weavers, metal workers, stone workers, leather makers, fiber makers, pearl workers, and the like—who formed small communities within the city, living and working together. As people developed specialties that were needed by the society they were given special time and space to flourish. Other family members might do their work and help provide food for their families. Perhaps at Cahokia these talented artists and artisans developed their own guilds and had special sections of shops where people came to trade for their special wares.

Trade and commerce were vital engines driving Cahokia's economy, so each artist and artisan was given the resources to develop. These products helped fuel Cahokia's capacity to trade with distant partners. Along with surplus foods and furs, Cahokians could offer fine fiber and cloth, distinctive pottery, basketry, and freshwater pearls. Also their stone works, especially pipes, must have been in great demand.

> **A cache of celts** was discovered in the summer of 2001 near O'Fallon, Illinois. Archaeologists are always digging (when they can), hoping to find more clues to past mysteries. "The Grossman Cache" held 70 finely worked celts (ungrooved axes) in a deep storage pit left by Cahokians 1000 years ago. The largest one ever found in this region weighed 25 pounds and was 18 inches long. This seems to suggest that toolmakers worked on communal projects. These celts were made of stone that originated in Southern Missouri, yet archaeologists think that these celts were made at Cahokia.

Fiber and Leather Makers Fiber making and leather tanning were ancient skills that everyone needed to know in order to survive. Over time some people naturally excelled at each of these tasks, making superior quality fiber and leather works. As communities of people settled together, these skilled craftsmen and women were very much in demand. They could stay with their special craftwork while other common folks did the basic labors of hunting, gathering, and farming. Then the commoners could have surplus, in good times, to trade for these fine craftworks.

Women often worked at the hide tanning in autumn and winter when gardening chores had been accomplished. Hide tanning was a time-consuming process, and women around Cahokia might have gathered in work stations away from the village center in order to visit and gossip together while they worked. Hide tanning is messy work and it is best done beside a stream or source of water.

Winter hides of deer, elk, rabbit, antelope, bison, and bear were best tanned with the hair on~as this was the thickest and best hair. Winter hides were also the toughest for moccasins and shields and sleeping robes. Autumn hides were best to scrape the hair from and fine tan for soft clothing. The same is true for spring and summer hides.

REFLECTIONS OF A FIBER MAKER

> This is the fictional first person account of a young man, perhaps 14 years old in Cahokia, who is skilled in harvesting plant fibers and twisting them into string and rope. This was a vital craft for thousands of years. Through his eyes we can learn something more about this age-old art and how some individuals and families may have developed these very special skills and trades. String and rope were necessary for fishnets, burden straps, house lashings, hunting traps and snares, as well as for weaving clothing and household accessories like sleeping mats. The mound builders even wove fine fabrics from these strong plant fibers.

I love the smell of plants in the sunshine after a big rain. I walk among the milkweeds in midsummer breathing the perfume from their flowers, and I watch the butterflies and bees at work. They remind me that my work with these plants will begin in the late fall and winter, during the moon of falling leaves. Then I will be so busy. I can see that these fiber plants are related—like cousins—to the dogbane, and butterfly weed.

Cahokia~City of the Sun 131

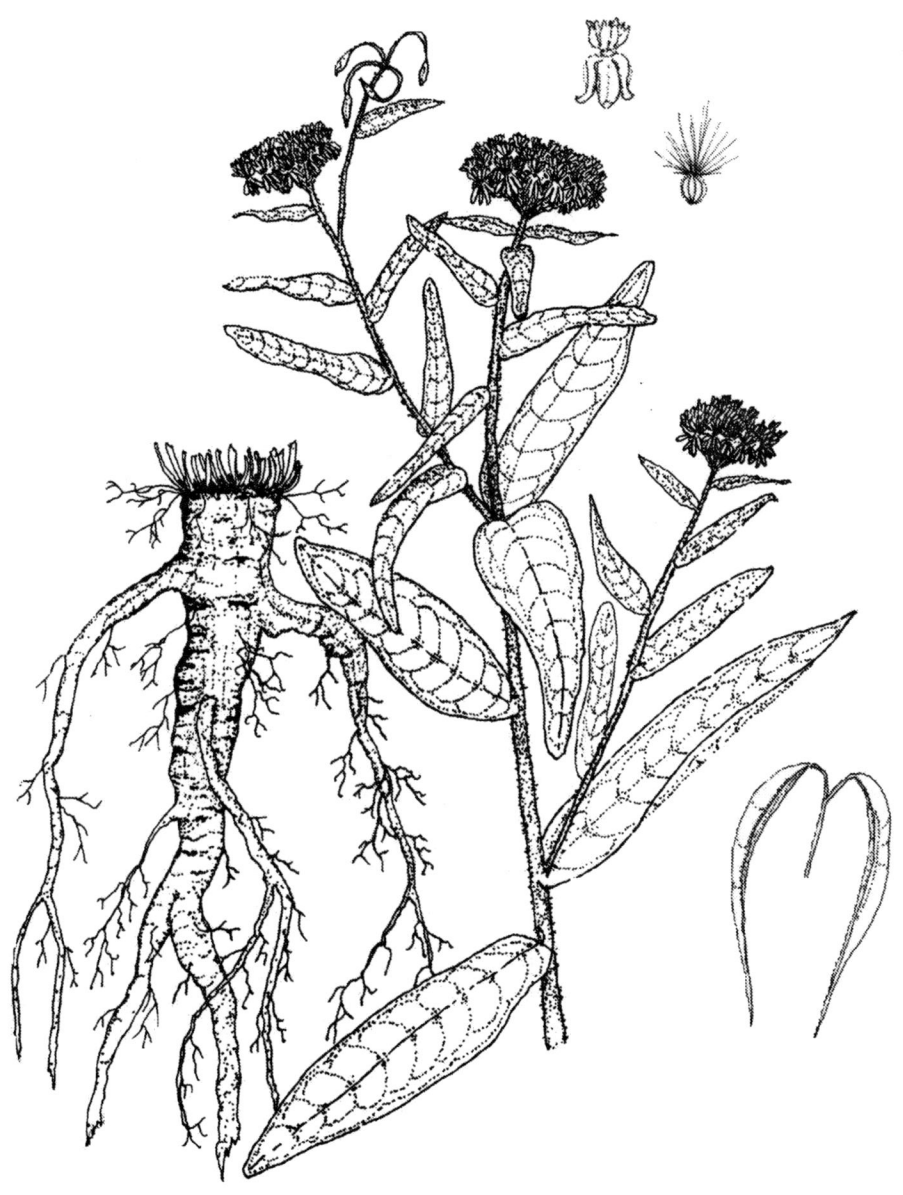

Butterfly Weed, *Asclepias tuberosa*, used for fiber making & strong heart medicines

My father taught me how to strip the thin skin of mature milkweed, butterfly weed, and dogbane stalks and peel them into fine threads to twist together. I roll these fibers on my thigh, twisting them into strong thread. Then I double or triple this thread until I can twist it into strong cordage. Some days I can make a great, continuous length of it and roll it in a big ball. My father and uncles weave this into fine fishnets and hunting bags.

I also make fiber and cord from the peeled inner bark of basswood, cottonwood, oak, ash, hemlock, and hickory trees. Sometimes, after a big lightning strike, I find trees that have been seared right open and their inner bark exposed. I take this and soak it to make heavy cords and rope to tie and lash burden baskets and house posts securely. I've also experimented with the bark of grapevines, and it twists into strong rope. The mature blade of some marsh grasses can also be twisted into fine twine.

My mother and aunts dye some of the fine cords with black walnut husks (black), and bloodroot (red), and white oak bark (yellow) dyes. This is then finger woven into handsome medicine bags. Even our chief, The Great Sun, has come to trade for our medicine bags. This is an honor for my family. My mother whispers to me that my father, a medicine man, once saved the great chief's life when the chief was having heart problems. He did this by making a strong tea of the milkweed and dogbane leaves and roots for him to drink. These plants are noted heart medicines. They also yield rot-resistant fibers that will last for many years.

A Fiber and Basketry Reference Guide:

Some plant fibers are far stronger and better than others. Early technologists knew the best fiber and basketry plants, and how and when to gather them.

Bittersweet, a perennial vine, has tough bark for fiber, cordage, and basketry.
Butterfly weed, a perennial wildflower, has fine fiber in its inner bark.
Cattail, a perennial marsh plant used for mats, padding, roofing, and much more.
Dogbane, a perennial meadow herb, with fine, rot-resistant fibers in its inner bark.
Dogwood, small trees and shrubs, with inner bark for fiber and basketry.
Grape, a perennial vine, with easily peeled bark for fiber, cordage, and basketry.

Hickory, a valuable nut tree, with tough inner bark for fiber making and lashing.
Maple, a valuable timber tree, with tough inner bark for fiber and basketry.
Milkweed, a perennial meadow herb, with fine, rot-resistant inner bark for fiber.
Nettle, a perennial herb, with fine inner fiber for twine and weaving.
Oak, a valuable timber tree, has tough inner bark for fiber and basketry.
Sedges and rushes, tall perennial plants whose dried leaves make fine cordage.
Willow, shrubs and trees whose inner bark makes fine fibers and baskets.

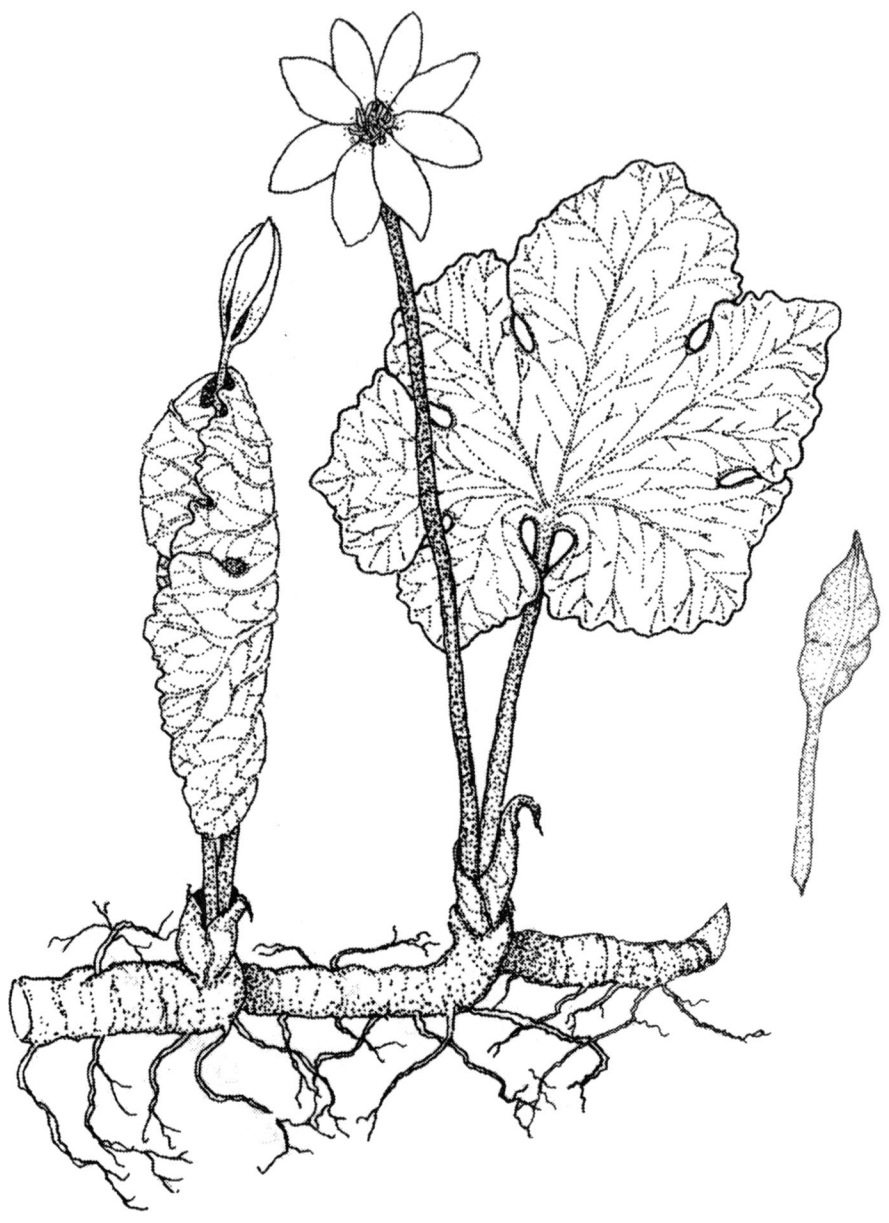

Bloodroot, *Sanguinaria canadensis,* used for dyes, insecticides, & skin care.

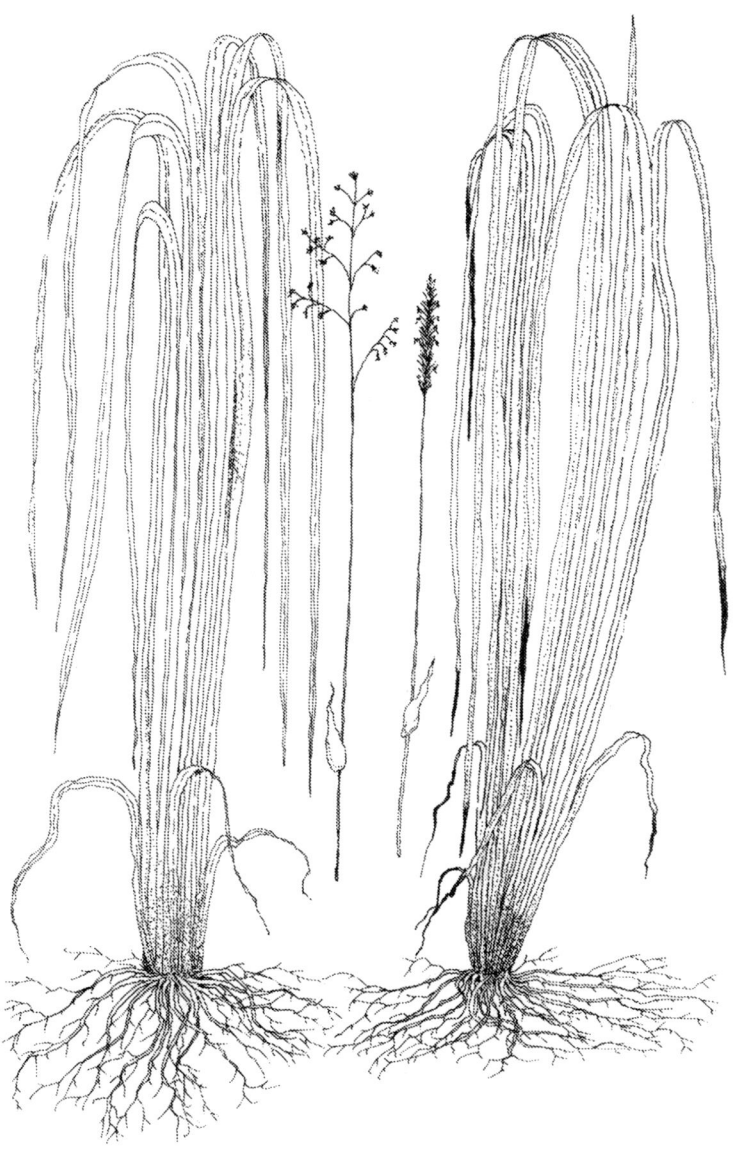

Sweetgrass, *Hierochloe odorata,* **& Sweet Vernal Grass,** *Anthoxanthum odoratum,* used for basketry & ceremonial needs.

Potters and Basket Makers

Pottery, basket making, and mat making are fine crafts easily learned and passed on, especially through families. These were important early skills that everyone needed to master in order to improve their lives. In time, skilled crafts people developed in each native village, and workers would come to them to trade their surplus foods and other crafts for fine pottery, mats, and baskets. Each of these crafts was time-consuming and required considerable attention to details. Both pottery and basket making were fine crafts in great demand among the mound builders, especially at Cahokia. Most Mississippian towns were built around large central plazas where craft people could assemble to barter and trade their craft-works for other items of need and beauty.

Potters first had to find and gather their clay from natural clay veins (deposits) in the earth. This had to be cleaned, sifted, and worked (much like kneading bread) in order to develop the clay's strength. Then special fibers and grit, like ground stone, sand, or powdered shells, were added and kneaded in to give the clay greater strength and stability. Shell-tempered pottery was a great innovation during the Mississippian period. Because of the added strength, pottery could take greater shapes and forms, plus have increased efficiency in cooking.

The potter could then begin to fashion pots and bowls when the clay was fully prepared. Otherwise, a finished pot might collapse or explode during the firing. Potters usually began making a pinch pot for the base of a clay jar; then pinch off balls of clay and roll them into long "snakes" and coil them on to the pinch pot, setting them with wet slip (like soft mud). This would be paddled, using a gourd fragment or flat piece of wood to make the pot's walls fairly equal in thickness. When each pot was finished, it was usually set in the shade to dry slowly for several days.

Firing the pottery required good weather and careful preparation. Most native clay products were probably fired in earthen, clay-lined firepits for several hours or half a day, in order to fully harden and finish the claywork. A pot then had to cool for hours before it could be handled.

Basket and mat making also required advance time to harvest and prepare the materials before weaving could begin. Many different vines, splints, plants, and inner bark could be used, but it first had to be gathered, prepared, and soaked until it was pliant enough to bend without breaking. Fine mats were woven from native river cane splints, and from cattail reeds and bulrushes, as well as some sedges and rushes. Baskets were also woven from these same materials.

Heavy workbaskets were made from maple, oak, or ash tree splints tightly plaited together. These were called pack baskets or burden baskets. Willow rods, hickory, and dogwood were also used, as were grape vines, bittersweet and Virginia creeper vines. Many of these created for ceremonial and ritual uses were dyed with natural colors created from nature.

Natural dyes were fermented or cooked from bloodroot (orange-red), poke (magenta), black walnut husks (warm browns), hickory nut husks (blacks), goldenrods (yellows and greens), sumac (array of browns), Oak galls (black), ripe grapes (blues), and a range of natural pigments. The mat and basket materials often had to be dyed in advance of weaving to produce lively patterns. Native People would sometimes soak basket materials in the dugout canoes, when not in use.

Cahokia began to decline in population and mound building activities after 1200.

Similar declines swept across other mound building centers, while some continued to flourish. No evidence of warfare or natural disasters seems to explain this slow decline. The soil must have become severely depleted from intensive farming and exhaustion of forest resources. There is some evidence and theories that diseases like tuberculosis and health malformations may have also been factors in Cahokia's decline. This great city was nonetheless inhabited until about 1400, after which it was abandoned.

Other temple mound building centers continued to thrive for several hundred years longer, well into the 1600s. Spanish accounts in the 16th century record fascinating encounters with several Great Suns presiding over remarkable villages, temples, and mounds south of Cahokia.

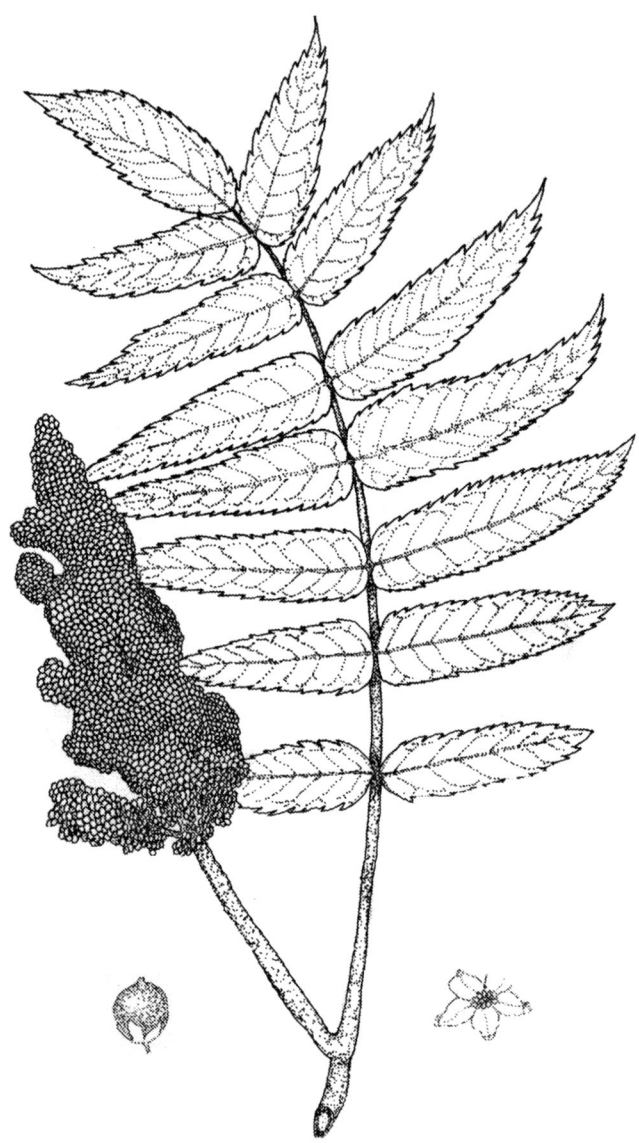

Smooth Sumac, *Rhus glabra*—a valuable perennial food, beverage, & dye plant

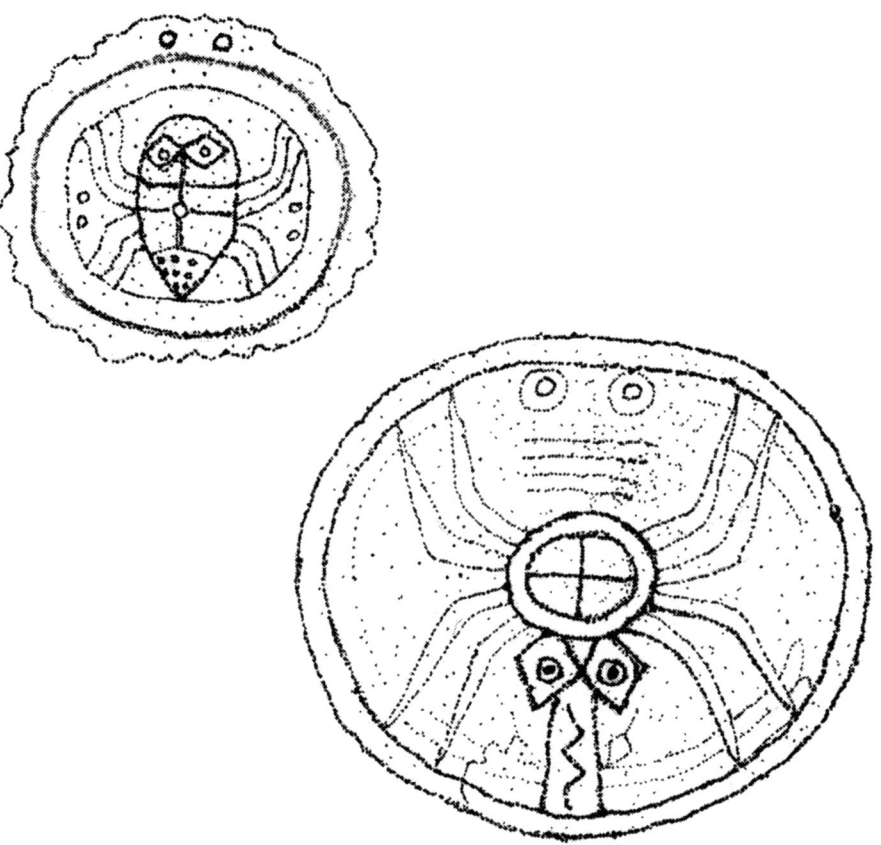

Spider Design Shell Gorgets–Southeast Ceremonial Complex
Possibly 1000 years old; very scarce. The Spider was sacred.

11

Daily Life~What was Daily Life Like for the Mound Builders?

o o o o o o o o o o o o o o o o o o o o o o o o o o o o o o

The Sun is my Father and the Earth is my Mother; on Her bosom I will rest.

—*Tecumseh, (1768?–1834) great visionary Shawnee Chief,*
1811

> What was life like for the mound builders that created so many amazing earthworks? Farmers, fishermen, hunters, gatherers, sachems, shamans, healers, herbalists, craftsmen—we look at some of these individuals, imagine their daily routines, and piece together their lives. Certain people had very special roles. Family life and the role of men, women, children, and elders were vital in the daily life of Cahokian society.

These memorable words of the great Shawnee Chief Tecumseh seem to encapsulate some of the basic religious beliefs of the mound builders, who created sacred landscapes, so many as breasts rising from Mother Earth. Most of the burials within were oriented to the east, toward the rising sun and the renewal of life itself. The sacred Sun Father would begin a new transmigration for the spirits ascending to higher dimensions where they would look down upon family, clan, village, and earthen cathedrals across sacred landscapes.

~~~~~~~~ Imagine a day in the life of a Mississippian boy ~~~~~~~~

Imagine a hot summer day in Cahokia in 1150. The Loon Clan families live near the big Twin Mounds, across the vast central plaza from The Great Temple Mound. Twenty of their homes encircle the west side of the conical mound, which holds the bones of their ancestors. Copses of wild plum bushes grow in an arc behind these homes. The Loon Clan People are privileged to live within the great palisade wall along with their distant Bear Clan associates, whose homes number about thirty and cluster around the nearby temple mound.

Both the Bear Clan and Loon Clan people also have ancient ties in the east. Their distant relatives once came from early mound building people six days running southeast of Cahokia on the Black Warrior River. They have lived in Cahokia for many generations now and are part of the elite leadership. These two clans have given Cahokia most of her healers and shamans, people "gifted with insight" and special ways of knowing things that are not normally clear.

Scioto, "Little Deer," has just walked down to the marketplace to help his mother and other Loon Clan relatives trade their fine river cane mats. He picked several ripe plums and ate them with relish. He is 10 summers old and a fine young athlete, although he slept too late to help carry the heavy rolled mats for his mother. She is never angry.

His mother Keewatin, "North Wind," and her sister Kansas, "South Wind," are beautiful young women. Their long black hair is neatly braided with loon feathers fastened above their ears. They each wear short, fringed deerskin skirts bound at the waist with fine fingerwoven sashes of milkweed fiber. Short, sleeveless deerskin tunics cover their torsos and hang almost to the waist. His aunt Kansas has a cute baby girl, who is sleeping bound snugly in her cradleboard. Kansas must pause occasionally to nurse her when she wakes up and cries. Scioto's mother lost a newborn baby girl last year. He would have liked a baby sister, but he would really rather have a little brother.

The marketplace is filled with noisy people and every kind of fragrance. Scioto's cane mats and baskets are traded away quickly. They are very much in demand. No one can match the Loon Clan men and women for their weaving skills. His mother and aunt have traded with a village potter for two big soup pots; they have also traded three baskets for two nice wooden bowls and wooden spoons. Finally his mother traded her last fine mat for a new flint hoe, hafted on to a strong hickory branch. Scioto has no time to play with his friends because he has to help carry these new items back to the house.

As they walked back through the palisade, several of his friends called out to him. He will meet them later and they'll go spear fishing for buffalofish in the lake. His father would want him to help the women now, until they release him to go fishing. His first duty is always to his family.

Scioto's home is a big rectangular house, entered through a narrow, low doorway between two large cedar posts. It is cool and dark inside and smells of mosses and sweet herbs. The dried mud walls are pale and decorated with some of his grandmother's fine river cane mats, plaited with geometric designs in red, yellow, and brown. The central feature is the firepit set in a hallowed-out bowl of clay and earth. The floor is hard earth swept clean daily with a hickory broom. A low clay altar behind the firepit holds the Loon Clan symbol, along with his father's stone bear pipe and smoking bag, and the sacred medicine bag.

The family's sleeping robes are rolled up against the back wall. Several nice clay pots and water bottles also stand near the fire and against the walls. A sheet of mica cut in the loon shape stands against the altar and reflects light. Each night the sleeping robes are rolled out around the fire. Grandmother Cataula, "Mulberry," spread her deerskin robe over cattail reeds near the fire. Scioto spread his deerskin robe near her on the earth floor. His mother and father spread their large buffalo robe in one corner of the room, where they kept each other warm. They actually kept only a small smoky fire during the cool summer nights, just to keep the swarms of insects away.

Grandmother Cataula has been sad since the body of her dear husband was cremated and placed to rest inside the Loon Clan Burial Mound five moons ago. His spirit has passed over into the great star realm. Grandpa had lived a remarkable life and was a cherished advisor and Medicine Man to the Great Sun. He knew more about medicines and healing than anyone else in the city. Scioto had been studying with him since early childhood and hopes to follow the healing pathway. He misses the special times when just the two of them would go fishing together for the day.

Scioto watches the little baby sleep and talks with his mother and aunt while they prepare the midday meal. This is the one main meal that they share together. They simply pick or snack on wild edibles or leftover dumplings if anyone gets hungry later. His mother revives a venison stew she made yesterday. She peels and adds more wild onions and wild garlic to the savory broth of corn and beans. This is simmered outside over the big campfire, which is burning low. The day is too hot for a big fire. Scioto stirs the stew with the cherrywood stirring paddle that his father made. They will scoop the stew out with a small gourd, and fill their gourd bowls. They use small mussel shells for spoons/forks/knives; one sharp mussel shell does everything. Later Grandma will take these few utensils to the creek to wash them in warm sand and water, while everyone else goes to work in the gardens.

Mahtowa, "Bear," Scioto's father has gone deer hunting with the Loon Clan men, because the deer have started invading the family gardens. The men have posted guards and have set several hunting blinds so that they can kill as many deer each night as possible. They hang the deer from the old oak trees near the garden to skin and gut them and cool them down. They hope this will offend and turn away the other deer from returning to the gardens. But it does not. Scioto's family certainly eats well, and has extra venison to trade in the market. He and his father have also taken half a deer to the Crow Clan People outside the palisade near the creek. Their men have been sick and unable to hunt since the last moon. They invite the Crow Clan women to return to the "butchering tree" with them and take all the venison they need.

Soon Scioto will go with the Loon Clan women to the gardens. It is his turn also to post guard and sit up high on the ramada in the center of the garden for three hours each day. He first gathers several pouches full of small rocks to make a pile beside him. He will hurl a rock and yell very loud each time crows or raccoons or rabbits come into the garden. This is the opportunity to practice his aim, too, and Scioto has a good eye for the moving target. He is eager to be able to go hunting with his father and the other men. He also takes time to help them with the butchering and distribution of meat. After his duties to his family and clan, he will be free to spend the rest of his day fishing with his young friends. Most of all, he is anxious to go swimming!

Lexicon of Names:

Scioto, Iroquois for "deer"
Keewatin, northern Algonquian for "north wind"
Kansas, tribal name for "people of the south wind"
Cataula, Creek for "mulberry"
Mahtowa, from the Sioux and Ojibwa for "bear"
Tecumseh, Shawnee for "panther"

The Home and Daily Life

We imagine that the common Cahokian home was a small round, square, or rectangular structure of log posts, sunk into the ground, with willow and alder rods woven between them and plastered over with mud, to make the house walls. This is often called "post-and-wattle" or post and daub construction, and is found in many places around the world from England to Africa to Mexico and Central America. A conical thatch roof was made of cattails and rushes around a center smoke hole through which smoke from the cooking fire escaped. A swept earthen floor was probably spread with large beds of mosses and grasses beneath sleeping robes of deerskin and buffalo hides, tanned with the fur on. The family might have stretched out close round the small central campfire for warmth on cold evenings and protection from insects during summer evenings. Cahokia's temples and houses for elite leaders were probably built along this same plan only much larger and with platforms for sitting and sleeping inside.

The typical house was little more than a comfortable sleeping shelter and place to cook and eat in bad weather. Basic life in most mound builders' settlements took place outdoors. You might have enjoyed one main meal as a family together around midday, and gathered wild edible "snacks" frequently at other times, like wild berries and nuts, wild onions, groundnuts, hog peanuts, seasonal mushrooms, evening primrose seeds, and sunflower seeds. Life was good and most seasons provided abundance for everyone.

Food preparations centered around the campfire, either within or just outside the home. Game meats were roasted on green willow or hickory branches, turned in spit fired or barbeque fashion as the branch was seated between two forked green hickory or cedar branches. These were driven into the ground so that they would stand upright perhaps six feet apart, with the fire burning between them. A large pot of soup or stew was probably always bubbling near the midday fire pit. Small fish were easily padded and packed in wed mud or clay and baked within the fire.

Each home must have had a wooden mortar and corn pounder, the mortar made from a hollowed tree trunk to a depth of about three feet; the pounder of smoothed, rounded oak or ash log was used inside the wooden mortar to literally pound nuts, beans, seeds, and corn into meal and flour. Many homes probably also had a stone metate and grinding stone that was used in the same way. Women and girls spent some time each day pounding or grinding coarse foods into meal, so that they would be more digestible when cooked. Cornmeal with beans, herbs, berries, or nuts, was patted into dumplings and dropped into hot soups, or wrapped in large grape or corn leaves and placed in the outer embers of the cooking fire to bake into "ash cakes." These were the ancestors of "journey cakes" or "Johnny cakes." Beechnuts, chestnuts, hickory nuts, acorns, wildrice, hazelnuts, and blueberries were favored seasonal fare, as well as seasonings, along with maple syrup in winter.

Evening Primrose, *Oenothera biennis*–valuable roots, leaves, & seeds for foods & medicines

The men and boys hunted the game and fish, and women often helped prepare it. The women and girls prepared the food and did the cooking. Specialties at Cahokia might have included: trout baked in clay, corn soup with wild mushrooms and corn dumplings, roast rack of venison, roast buffalo, wild turkey stew; seasonal favorites might have included frogs legs and wild onions, boiled beavertail, roast duck and Canada goose, baked catfish with blueberries and huckleberries, sunflower seed and cornmeal ash cakes. Jerusalem artichoke root soup and "Three Sisters" stew of corn, squash, and beans with dumplings were doubtless village favorites.

The Marketplace

Much of everyday life in Cahokia almost 900 years ago must have centered in the marketplace in the western part of the city near the waterfront. During the early part of the day, Cahokia's gardeners gathered to trade their surplus produce and barter for things they needed. Women would trade excess sweet corn, strawberries, purslane, wild rice, and fresh or fried squash and pumpkin for cordage, or pearls, or a new hoe. Young children might trade small buckets of fresh blueberries or large sacks of wild mushrooms for a small spear or bow and arrows or perhaps a new doll. Men brought surplus meat and fish and handmade craft items like wooden bowls and spoons to trade for fishnets or burden baskets or special items that they could not make for themselves. Artists and artisans brought their finest wares to trade for food and other special items.

The marketplace was usually a beehive of noise and activities. Situated near the inland watercourses, this was the first repository of Cahokia's wealth of trade items. Of course the Great Sun was always given first choice of everything displayed here. His young warriors would bear him into the marketplace on his golden litter, with much fanfare on the days he chose to visit. Teams of Sun Priests went before him to prepare the way, following a team of drummers and chanters who would sing the honoring songs. His bearers would stop wherever he called out, and things were lifted up for him to see and inspect items that pleased him. He would often distribute small copper ornaments bearing the Great Sun symbol, the great spiral encircled by the sun shooting out rays of light all around the circle. He traded one of these precious emblems for goods he wished to take from his followers and the other traders.

Once each month there was additional excitement. On the three days of the full moon traders would come with exotic items and their own unique pottery, stoneworks, and bark buckets from some of Cahokia's satellite villages. These were amazing days of exchanges of stories, ideas, commerce, and gossip. Sometimes by prior arrangement, a young man or maiden might return with the distant traders to train in another field of artwork, or as a marriage partner. Cahokians were forbidden to marry within their birth clans and would often choose to leave for another village and a new life. The Great Sun always sanctioned these exchanges because they wove the villages together with stronger bonds.

Many dozens of specialized workshops were scattered throughout Cahokia where talented artisans created fiber, cordage, and woven materials, and fine hammered metal ornaments, and toolmakers crafted the implements for gardening and other work. Potters worked with native clays, jewelers created fine items for adornment, hide dressers tanned a variety of native animal hides and furs for use, and basket weavers prepared fine food-sifting baskets as well as large, tightly woven burden baskets that the laborers would carry on their backs, filled with earth, to increase the size of Cahokia's many mounds—daily. Many of these specialized craftworkers were centered along one aisle of Cahokia's marketplace, where these artists had special stalls along a grassy knoll. This was a place of distinction for Cahokia's finest talents.

Later in the morning people would turn to the day's work—hunting, gardening, farming, mound building or specialized craftwork. Some people remained in the marketplace all day, especially the artists and artisans, and trade visitors. Most Cahokian women and girls went as early as possible to work in their gardens, and gather what seasonal produce they could for the next market day. Babies and toddlers stayed with their mother and the mother's extended family, who looked after them by turns while also doing their work. Most Cahokian men and boys were required to spend many days transporting fresh earth to the mounds and stamping it down evenly. Rainy days, or at least one day each week, the men went hunting and fishing to help supply the cities tables.

Hunting and Gathering

Cahokians depended upon deer, wild turkeys, beaver, and rabbits for food, along with a broad range of fish and freshwater shellfish. During seasons of plenty life was fairly easy and there was probably plenty of free time for both the men who hunted and fished and the women who gathered wild plants and gardened.

Shellfish must have been a major food source judging by the abundance of freshwater pearls found in Cahokian burials. Artisans made many thousands of beautiful beads from the shells. Women and children probably hunted the mudflats year round for the mussels, which must have generously filled woven net bags and bark buckets. Men went after the game animals. They hunted buffalo in summer and fall, and deer all year.

When game was scarce, the tribe or clan would hold a hunting ceremony to add special blessings to their efforts on the evening before the hunt. At Cahokia the individual clans might have each sponsored such events, or grouped together to collectively offer prayers and good energy toward favorable hunting. The shaman would go into trance and journey to find where the game animals were waiting for the hunters and return with this mystical advice to insure success.

Men and older boys left on rainy days to hunt and fish, when mound building was usually discontinued. It was probably a custom to bathe early and dress lightly; then bathe in the smoky smudge of herbs in order to mask the human scent. Each woman usually braided her man's and son's hair (sometimes intertwining it with sweetgrass). They would wear very little other than a breechcloth, just a quiver of arrows slung across the back and they carried a bow. Some men would wear a bandolier bag, or medicine bag, strung across the chest. This would hold hunting medicines, personal totems, hunting charms, and perhaps some parched sweet corn and sunflower seeds for energy. Others might simply wear a small leather bag filled with pemmican (dried, concentrated energy food made from animal fat, nuts and berries.) Hunters were sometimes gone for days before they returned laden with fresh meat.

Many times hunters would go together in hunting parties because much of their prized game like deer, buffalo, wild turkey, and Canada geese traveled in herds. It was easy to bag 10 or 15 or more at a time, and then set up a temporary hunting camp to process the meat and skins. Sometimes the women might join them and help with this processing, but usually the men did this themselves and returned to the village with just the essential meat and best hides.

Division of labor was vital in early Indian communities, especially Cahokia, where all of the people were expected to spend some time working to build the mounds managed by the subchiefs and surveyors. This was an ongoing task important to the whole city.

Farming

Cahokia women and girls worked in their fields outside the central city and beyond the palisade. While the men and older children helped to prepare the fields, and sometimes accompanied the women, farming was principally women's work and the women owned their fields. It was a Native American worldview that women controlled the fertility of nature and could produce fertility~both in the families and in the fields. Women were considered the more auspicious gardeners because of this outlook. Young children accompanied their mothers in the fields, and as soon as they were old enough, they were given simple tasks, like tossing stones to chase birds and animals out of the growing fields before these intruders could steal seeds or any produce.

Men were the toolmakers who fashioned sturdy hoes for their women to use in the fields. Women (and also men) wove and plaited workbaskets used to carry field produce, and sifting baskets to help in the processing raw corn and seeds before the food was cooked. Women often accompanied men on trading expeditions to the south in order to trade for new varieties of seeds and learn more about planting and fertilizing techniques.

Mound Building

Each day during favorable weather large teams of laborers streamed across Cahokia carrying 50 to 60 pounds of earth each in burden baskets on their backs to build up the various mounds in this great city of mounds. This was the most essential work of Cahokia and everyone had to help perform these tasks. Usually the nobles, artists, and artisans were exempt from this heavy labor.

The physical structures of the mounds must have symbolized the Mississippian ideological world. The mounds must have represented powerful conduits through which power, prayers, and sacred information flowed between the three structural layers of the Mississippian world: the upper world, the middle world, and the lower world, plus the many elaborate layers and levels in between. Dirt was brought from the lower world to the middle world of the living, and piled up high to reach the upper world (over and around the burials) to improve the elites' abilities to communicate with each of the worlds.

Burials within these immense breasts, the conical mounds, on Mother Earth usually faced east to the rising sun. Perhaps the temple mounds were also perceived as wombs from which the spirits of entombed leaders would rise again. Sacred accoutrements accompanied the burials enabling safe, mystical journeys to the vast spirit world where the great leaders would palaver about their returns and how to satisfy the peoples' needs.

The mound builders must have chanted work (or sacred) songs as they tramped along these ancient footpaths to build each mysterious, unforgetable mound. Village overseers and engineers would have directed the laborers. The conical burial mounds must have taken on amazing energies over time from all the spirits surrounding them, while living spirits were working on building them higher and higher. Some of the conical mounds may have represented leaders in the main clans, who lived in the areas around each mound holding their beloved ancestors within earthen breasts rising ever skyward toward the Creator and the Spirit realm.

Mound building was the central focus in Cahokia for many years, yet by 1150 this work had reached a peak and began to taper off. So many mounds had been well built. There seemed to be some ceremonial shifts to greater ritual performance on the mounds, rather than continuing to build them up. It is unclear what caused this. But perhaps this was just the next phase of life evolving naturally. Perhaps Cahokia's burgeoning population simply required more attention to survival necessities and developing food surpluses, trading commodities, and protection for a burgeoning society.

The Role of Children

Children were, for the most part, cherished in traditional mound builders' societies. They were given considerable freedom, and nurtured within the clan systems into which they were born, which were like extended families. Children were usually members of their mother's clan, while their father's clan was also of strong help and interest to them throughout their lives. The descent line passed through the mother's family, much as it does in American Indian tribes today.

Children were challenged to play many games of skill and chance to help develop their coordination. Running games, relay races, spear throwing contests, and bow and arrow contests helped test their natural abilities. Chunky was perhaps a most favorite running and spearing game for most boys and girls. A large round, disk like chunky stone about the size of a softball was rolled along the ground while the chasing marksmen threw spears at it to stop it or mark where it would probably stop rolling. Many prehistoric chunky stones have been found throughout the mound builders' regions and further south. This was also a favorite game among the mound builders descendants, the Creeks and Choctaws in the Deep South.

Children's dolls and games were often given in preparation for the roles in life they would grow into, and served to focus their attention on natural goals, like hunting, fishing, child rearing, and cooking. Children began helping out around the home at their earliest ages. Their work habits were welcomed and complimented. Parents, siblings, aunts and uncles watched each child to see if they showed signs of a certain destiny or passion, like artistry or leadership skills. This would invite some special grooming by the watchful clan relatives.

Grandparents told special stories around the evening fire, which were the teaching tools of mound building culture. Stories explained how the world began, and where people and game animals came from, and why storms were often dark and menacing. Stories also served to teach the right behavior.

Drums and rattles, music and dance were valuable aspects of mound builders' life that children enjoyed. Children were groomed for their life path through ceremonies and with songs. The following story illustrates some of this.

A CAHOKIA CHILDHOOD

> What did it feel like growing up in Cahokia? This fictional visit attempts to recreate an 8-year-old's life in Cahokia almost 900 years ago. The sidebar introduces some Indian names and meanings.

I was born during the Moon of Shedding Feathers when the marsh grasses were high. My mother Waco was working in the cornfields and could not get back to our clan's birthing hut in time, so I was born in the tall marsh grasses and fragrant sweet flags. This is why she called me Waukau, because she gave birth to me alone in the flags. Perhaps this is why I love this region so much. I like to stretch out in the sweet flags, so fragrant in the hot sun. Then I smell just like them.

My mother Waco was named for our Heron Clan, because her mother (my grandmother) was the clan leader and a noted herbalist and healer. My father Waunakee is the Peace Chief for his White Water Clan. Both of my parents are important leaders and subchiefs in Cahokia. "Women are strong leaders," my father says. "They will make our future bright and strong. Women chose the men who will lead."

My little brother Tamaqua is just out of his cradleboard and beginning to walk. I help my mother look after him and he is a handful! He is really a busy "little beaver!" I love him dearly. He is almost more fun than playing with my doll.

When my mother was pregnant with him, she made me the most beautiful doll of soft deerskin stuffed with dried grass and sweet flags. She wears a lovely dress of rabbit skin and a bead on a sinew cord for a necklace. She has long cornsilk braids that are black like mine. I call her Menahga because she helps me find blueberries.

My parents are concerned now and talk softly together into the night about our safety. They say that some of the subchiefs want to build a great palisade around the core of Cahokia to protect the Great Sun, our great chief, and the burial mounds of our ancestors, as well as our many Cahokians. Some of the traders are saying the Southern Cult, or Death Cult, is planning to make trouble in this region. I don't think it's true. But, my mother says I am just a child, and must not worry about such things.

We will go tomorrow to take cornmeal offerings to the great burial mounds. This is the beginning of the tribe's fall equinox celebrations. The subchiefs and drummers will gather up on top of the Great Temple Mound. The Great Sun will lead us in prayers and dancing, and then we will cook and share a great venison feast together. The women will all make my mother's tasty recipe for Three Sisters Stew, using the corn, beans, and squash from our many fine gardens. There will be thousands of people to feed.

Many people are coming here from the subtribes that live well beyond our city. We will have many days of trading and honoring visitors, ancestors, and the harvests. Waco says we will take our sleeping robes and all sleep at the Great Temple Mound. I love to feel its powerful energies. I have great dreams when I sleep there. The subtribes will congregate and sleep at the base of the Great Temple. Children are never allowed to climb or play on its steep slopes. It is too difficult for us to climb the steep temple steps, and besides it makes me dizzy to look up there. The chanting and fragrance from the sacred temple fires drift down around us and it is so exciting.

Last week I made my first pots with my aunt. She is teaching me how to create fine pots like the ones she is famous for. All of the traders want her pots. She taught me how to work and knead the gray river clay until it is smooth; then add fine powdered mussel shells, and roll out long, fat strips with my hands. These coils are carefully added, one on top of the other, fitting in with wet clay slip. My hands are quick to smooth the coils into solid clay skin using an old piece of smooth broken gourd. I would like to become a potter, like my aunt. I could just work in clay all day. I make some little beads for myself and for my doll. I even love to go along and help the men and women dig the clay from the various raw clay veins above the Great River.

We left our pots drying in the shade of a hickory nut tree. Today, when I go back to my aunt's place, we will build a big earthen fire and slowly move our pots into it with long sticks, so they can be baked and hardened. I am finishing an offering pot for my mother to hold her sacred cornmeal.

Yesterday my father returned from a long trading journey to Appomattox, far to the east. He and the other runners brought back fine tobacco seed, dried tobacco leaves, and different kinds of corn and bean seeds for our gardens. Our farmers will plant the seeds during the spring planting moons, after they have been blessed in our kivas.

My grandfather Kokomo is one of the best Sky Watchers. He helps the high council advise the Great Sun about when to plan each ceremony. Grandfather also helps direct the many hundreds of mound builders, who must work daily to haul earth from the lowland barrow pits in big burden baskets on their backs. Each worker carries about 70 pounds of earth up the steep slope of the giant mound and empties it on the top.

The men in our Loon Clan weave the strong burden straps used on each basket to encircle the men's shoulders or foreheads. The Crow Clan men weave the heavy-duty pack baskets out of willow, dogwood, and basswood strips tightly plaited and secured with deer sinew. They need to work almost constantly to keep up with the needs.

This great mound has been growing for many years—since before I was born. They only work on good days. It is too dangerous in the rain. They are happy to contribute so much energy to these great monuments. Our best foods go to the mound builders because they must be happy and well fed in order to do good work.

NATIVE NAMES

Waco is a Muskogean tribal name meaning "heron;"
Waunakee is Algonquian meaning "he has peace;"
Waukau is Algonquian for "sweet flag;"
Tamaqua is Delaware for "beaver."
Menahga is Ojibwa for "blueberry."
Appomattox is eastern Algonquian for "tobacco plant country."
Kivas are underground circular prayer rooms, like churches.
Kokomo, after a Miami chief's name, means "black walnut."

Herbalists and Healers

Herbalists are people skilled in the knowledge and use of herbs. They were societies' first scientists and doctors, because they could often treat people's illnesses, cure fevers, rashes, wounds, and assist in childbirth rituals. Herbalists studied all through life to learn about the many uses of wild plants. In some southern societies they were called curanderos or curanderas, medicine people. Each specialist might know the uses of two to three hundred herbs in their lifetime.

Healers are people skilled in the abilities to heal people, either with herbs or just their own special energies, or a combination of things. A few people were healers, herbalists, and shamans, but not many were usually that talented. Each individual held a special role in tribal societies. Often healers and herbalists would work together with the shamans to heal some patients.

Native medicine men and medicine women were also the herbalist, healers, shamans, and special mediators for their people. They would appeal to the spirit world and the spirits of sickness in order to help drive away illness. Each medicine person would amass a personal medicine bag, or bundle, filled with secret and sacred items to help their work. Along with special roots and herbs, paints, spirit objects, fetishes, and amulets (items to bring good luck), a medicine person might also have a whistle and certain healing songs. Drums and rattles were also used to perfect the healing and send evil spirits and illness away. A medicine bundle would be one of the most sacred items in the tribe. Some tribes held special renewal ceremonies each year to bless the medicine bundles.

Early societies prized their herbalists, medicine people, shamans, and healers, who specialized in these unique fields and often trained others to follow in their footsteps. Indians who suffered a serious illness and were healed by such individuals, like the little girl in the shaman's dream, were then called into the specialty, like a profession. They might devote their lives to learning about herbs, healing, and shamanism.

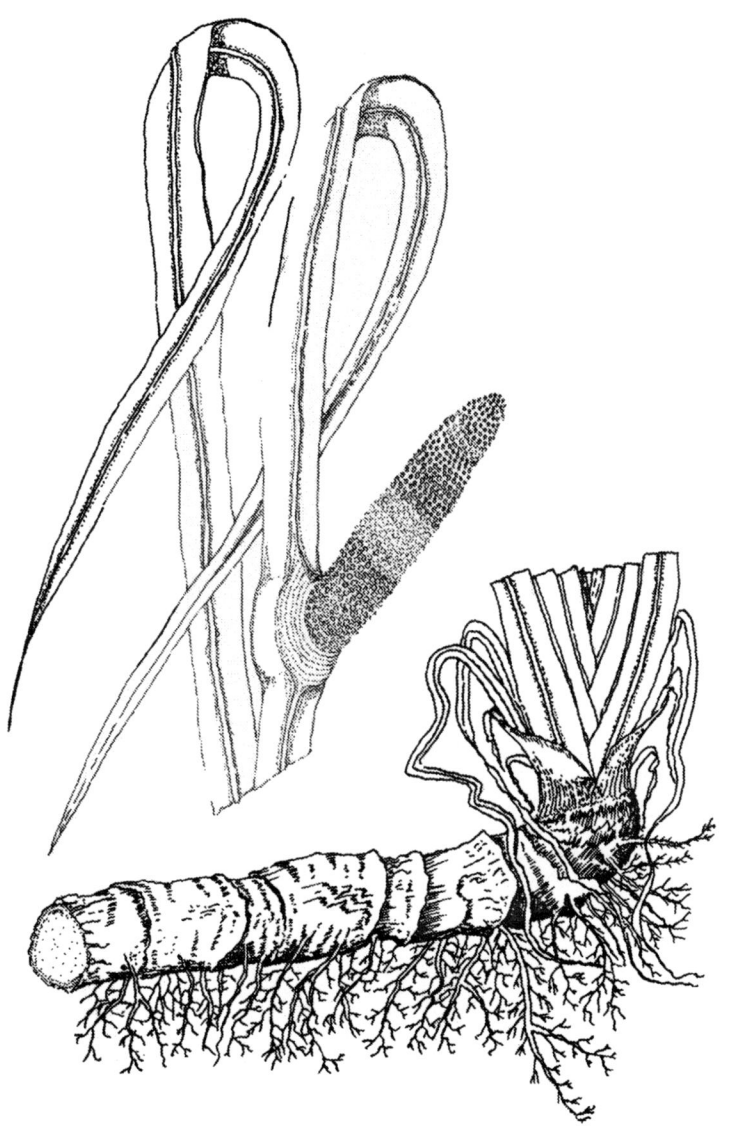

Sweet Flag, *Acorus calamus,* valuable perennial medicine & ceremonial herb usually found growing in large colonies in wet areas. It is highly aromatic.

Native Medicines and Healing Practices

The mound builders knew the uses of hundreds of wild plants. Many individuals had to know the simple healing herbs, because their lives depended upon this knowledge. Indian mothers and fathers knew that strawberry leaves steeped in water would relieve diarrhea when chewed or drunk in warm water, and calm insect and spider bites, and soothe skin rashes when washed on the skin. Raspberry leaves steeped in water could aid digestion and calm an upset stomach. Goldenrod and bee balm leaves dried and burned in a smoky smudge would aid breathing difficulties and treat colds and headache. Wild cherry bark steeped in water would relieve coughs, colds and sore throats.

The runners and hunters knew the value of using wild lettuce leaves and roots against snakebite, and the milky sap on warts and bee stings. Sunflower seeds and leaves, and evening primrose seeds and roots were eaten as endurance foods, especially along with prairie coneflower, and Joe Pye weed. Sunchoke (Jerusalem artichoke) and goosefoot were good wildfoods along the trails, and eating the wild mints, wild leeks, and wild onions helped digestion and kept insects away.

As native cultures evolved, they established healing societies. The Bear Clan was usually made of healers in many Indian societies. Yet there could be healers, herbalists, and shamans in every clan. Cahokians may also have had secret masked societies noted for healing capabilities and special rituals.

Many additional healing plants were undoubtedly learned about and traded during the mound builders' extensive trading excursions. The diffusion of healing knowledge reached far and wide during ancient times. We continue to rely on some of their ancient wisdom. *Sassafras* is an Algonquian word for "green twig"—and these green twigs were enjoyable chew sticks that also served as valuable toothbrushes. The mound builders must have known of hundreds of beneficial plant stems and twigs to use to clean their teeth.

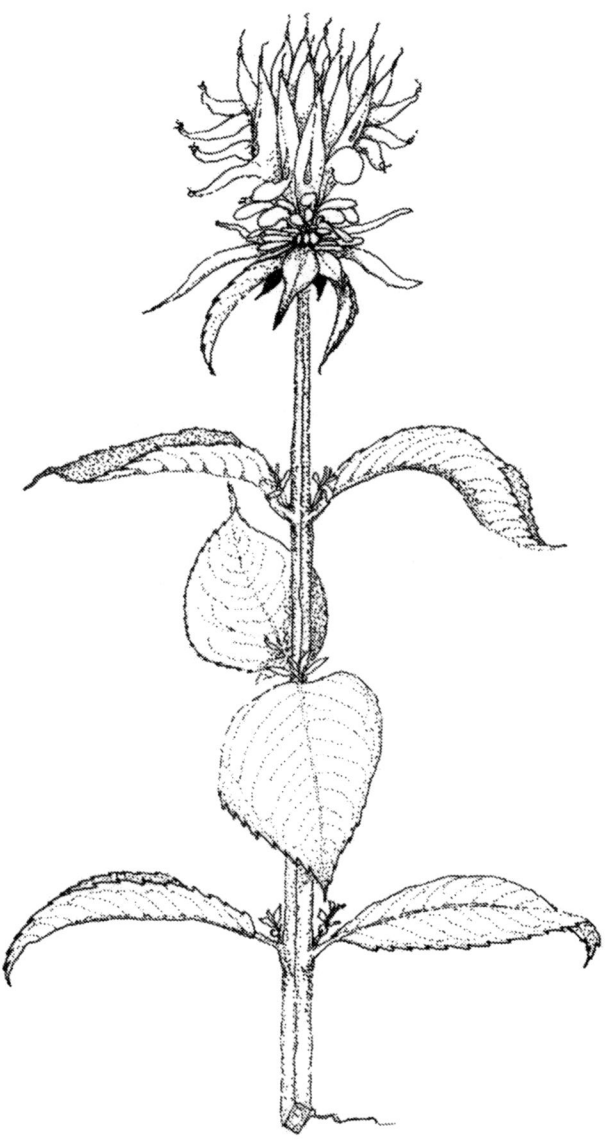

Bee Balm, *Monarda didyma,* a valuable wild mint for food, flavoring, perfume, beverage & medicine. Highly fragrant & attractive often growing in large groups.

Modern medicine has drawn on traditional American Indian herbal knowledge for dealing with many health concerns. Naturopaths, herbalists, and homeopaths rely to some extent on native medicinal herbs.

Cohosh is an Algonquian word for "healing roots." There are four species of cohosh: Black Cohosh, *Cimicifuga racemosa,* Blue Cohosh, *Caulophyllum thalictroides,* Red Cohosh, *Actaea spicata,* and White Cohosh, *Actaea alba,* or "Dolls Eyes." Although toxic to poisonous, these plants were used for millennia to treat an array of native health problems. Today both the Blue and Black Cohosh are used for the benefits of estrogen and other valuable alkaloids that can help men and women regulate their body's needs during different phases of development. Women use these herbs, with caution, during pregnancy and childbirth, and later in life during menopause.

Kinnikinnik is a Cree Indian (Canadian) word for "botanical mixture" that was usually made up of sumac, dogwood, bearberry, goldenrods, asters, wild mints, sage, and tobacco. This was used for incense, smoking, praying, and in teas to sooth and relax stomach problems. *Saskatoon* is also Canadian Cree for a particular small tree, *Amelanchier canadensis,* whose bark, leaves, fruits, and seeds were used for many different foods and medicines. During historic times the ripening Saskatoon berries would help to set the time for many Plains Indians' Sun Dances. *Tamarack* is another Canadian Cree word for one or two trees: the American Larch, *Larix americana,* and Black Ridge-pole Pine, *Pinus murrayana.* Each of these were used for many technical needs, and their inner bark and roots were used in healing. Today an extract of *Larix* is used in cancer treatments.

These Native Indian words may have their roots in ancient mound builders language, as the knowledge of their use is ancient. Today we rely on the *Oxford English Dictionary* to tell us the origins of words, yet this source can only rely on the first written and spoken accounts of each word. No one was here to hear the mound builders speak their language, which must have influenced the language and word uses of many North American tribes because of their great trading authority.

Dreams and Visions

Dreams and visions have always been given considerable importance in Native American societies. These were (and are) two strong pathways into the future. Through dreams and visions individuals could explore the unknown, and the spirit could travel to places beyond where the body could go. Dreaming is closely akin to shamanism, which generally required training and development.

We can only imagine about the mound builders' dreams and visions. Perhaps some of their mysterious artwork evolved from dreams, especially the ghostly Southern Cult artwork with its otherworldly symbolism. The striking "weeping eye" and stylized spider motifs, the Ivory-billed woodpecker and falcon designs radiate environmental strength and leave haunting impressions. We do know that Native American traditions attach considerable importance to dreams, especially children's dreams.

> Dreams happen naturally. Everyone dreams every night, yet not everyone remembers his or her dreams. We are always given special information in our dreams. Sometimes we see special artwork that we are to create, or a story that we must tell. Dreams can often foretell coming events, even illness. Dreams seem to come from a magical realm that we do not personally control.
>
> Visions are like daydreams that might come at any time. You get a sudden instinct about someone or something. You don't control this ability, exactly, but many people can block it, so as not to receive this information. Some people are natural visionaries.

Native dreamers were often considered visionaries and "seers" because they could forewarn the people of coming disaster, or tell hunters where to find game, where to fish, and where the seeds should be planted for better garden produce.

Many Indian youths would seek a vision at the time of their puberty, their coming of age. They would undertake a vision quest alone in a remote place, usually with the help of a medicine man or spiritual leader, who would train them for the ordeal and get them ready to fast, pray, and concentrate on receiving a vision in an altered state of consciousness.

A Healer's Dream

> This is the fictional first person account of an older Cahokian man, a healer, who is concerned because his wife has a tumor. He could be 40 years old. He has recently experienced a dream bearing information about where he must go to gather healing medicines, and then how he is to use them to help his wife

A strong presence came into my dream to show me where to find the healing plants I must gather for my wife's tumor. I felt a shiver of recognition as the ancient Medicine Spirit walked with me over to the old oak tree that stood near Lizard Mound. I looked around at the Thunderbird and the other prominent mounds that surrounded me here. Certainly medicines taken from this area would be very powerful!

Then I saw the large, dark knob growing out of the tree's upper trunk. This great burl looked like a tumor on the oak tree. Yet, the tree was fine and healthy. This reminded me of my younger days, when I felt that I could cure anything.

Medicine Spirit taught me how to sing and pray to the tree's spirit and made a tobacco offering to it; I watched how he scraped just enough bark from the north side of the burl for the healing treatment. "This will not kill the tree," he said. I watched him carefully, realizing that I would soon need to make this long journey and I must do everything just right. I turned to look at the giant earthen animals again. Seeing them covered with snow was a chilling sight!

Shamans and Sachems, People who Rule

Certain men and women in each tribe were gifted with special powers, like the ability to "see into the future" and communicate with the ancestors spirits, or even determine the cause of illness and seek its cures, and locate the game animals, or an enemy tribe. These individuals were visionaries and shamans. Shamans were the first scientists and priests. They could assist the herbalists and medicine people in developing healing practices. Their spirits possessed the ability to journey into non-ordinary realities and experience out-of-body travel. They had spirit guardians and power animals that helped them in the other worlds. Shamans were leaders of spiritual and ceremonial rites.

Sachem is an old Algonquian word meaning "leader or ruler." Men and women sachems (women were sometimes called *squasachems* or *sunksquas*) were the political elite in mound builders' societies, as were the shamans. Sometimes a person could possess skills both as a shaman and sachem. This was a very powerful individual.

The people in each village looked to their leaders to protect them from enemies, food shortages, and wild weather conditions, illness, or even bad dreams. Great importance was given to dreams and it was widely believed that nothing happened in reality until it had first been dreamed. People took great stock in their dreams. Shamans had especially big dreams. They also possessed the abilities to help others understand their dreams and act out the fuller meaning of dream actions.

Children were especially encouraged to share their dreams each morning. Indians believed that special insights and new information often came through to children first, because they were perceived to be pure.

A Shaman's Dream

> This fictional account of a Cahokian woman shaman reveals what happens as she awakens from a prophetic dream that has upset her. She may be 25 years old. She needs to think about her dream without disturbing her sleeping family. Dreams were vital pathways of learning outside of normal everyday ways.

She awakened suddenly from a big dream, and got up silently to go outside and think, while her family lay sleeping. She had seen the face of a village child sick with fever and covered with an irritating rash. She knew this little girl; she had helped her mother, as a midwife, give birth to her seven years ago. In this dream the child was desperately ill. As the village shaman, she would be expected to divine the cause of this illness.

Dream guides had taken her into the woodlands to show her strange new plants that she must use. She would be expected to teach the herbalists about these herbs after she worked on the infected child. There was a sense of urgency and renewal.

The spirits of her ancestors had come to her singing and waving wild cherry branches...They spoke in the old language about healing formulas of various herbs that she must now begin to gather from the land. She must dig and pound the pokeroot to a pulp and place it in a pot of water to ferment. She should scrape the wild cherry bark downward from the south side of the tree until she gathered three hand-

fuls. This would be added to the pokeroot, along with wild bee balm, raspberry, strawberry, and woodland violet leaves. She must send the runners into the eastern woodlands to bring back witch hazel twigs and leaves for this formula.

She began to softly sing, going into a meditative trance in order to journey back into the dream. There was more that she needed to know about the healing herbs in making the formulas correctly and how long to use it. Her power animals came running to her; she merged with them and they flew back into the dream.

Her ancestors' spirits came dancing to her again, bearing more herbs and reassuring directions about their use. She gained the vital details, along with the awareness that the little girl would take sick in five days and be brought to her on the sixth morning. She would be prepared to treat her with strong healing herbs, her drum and rattles, whistle and fetishes, and drive the sickness away! She would make her a curing fetish, an amulet for safety and good health, in the shape of a small bear.

In time this little girl would return to study healing and shamanism with her. Their friendship and shared knowledge would save the tribe many times.

Security

The heart of the commercial and ceremonial center of Cahokia was well guarded by elite young warriors, whose duties called for them to be vigilant about intruders and watchful to prevent any disruptive behavior among Cahokians. Perhaps they served much like policemen—maintaining everyday law and order.

Other groups of Cahokian warriors served the elite leaders, and were probably much like the Marines—an elite fighting group responsible for maintaining ultimate peace and safety. These warriors probably wore special chest plates and gorgets as symbols of their rank. They may have carried war clubs or worn them dangling from a belt around their waist. Bows and arrows were carried, too, and perhaps obsidian or flint knives.

Cahokia had grown rich and powerful and must have faced the realities of regional jealousies from outside its district. More temple mound sites were erecting palisades around their central core for protection. Cahokia must have been wise to these problems. The irony is that impressive warriors and palisades could not keep out disease or prevent internal strife and problems.

> The Spanish recorded in the 1500s how impressed they were by the powers of the Great Suns who ruled several centers where they stopped further south of Cahokia. They especially remarked about the numbers and skills of the warriors that the Great Suns could call upon.

12

Celebrations and Sacred Rites~Ceremonial & Ritual Possibilities

> *We give thanks to the Earth, and we give thanks also to all the things it contains. Moreover, we give thanks to the visible sky. We give thanks to the orb of light that daily goes on its course during the daytime. We give our thanks nightly also to the light orb that pursues its course during the night. So now we give thanks also to those persons, the Thunderers, who bring the rains. Also we give thanks to the servants of the Master of Life, who protect and watch over us day by day and night by night.*
>
> —From a Seneca Corn Dance address in autumn.

Temple Mound Builders' celebrations of life, death, the solstices, and the hunt are explored here, as we imagine more about these fascinating people, and how and why they created and used their mounds. The use of sacrifices may show some links with the Maya and Aztec, who were probably southern trading partners.

Celebrations of life

Mound Builders certainly celebrated Mother Earth and her multitudes of natural resources. Life depended upon secure knowledge of the natural resources and mound builders' populations increased steadily through time. They had evolved from hunting and gathering cultures with deep knowledge of the survival arts. The rulers of Cahokia may have belonged to the Dhegihan Sioux, who once included the Kansa, Omaha, Osage, Ponca, and Quapaw (Arkansa), according to scholars who have studied their oral traditions and historical evidence. Perhaps through vast trading networks and intermarriage with trading partners, mound builders represented many native bloodlines.

Life was celebrated and cherished with fine ornaments to adorn the body. The mound builders created a magical ceremonial world wherein their rites and rituals wove the years together. People celebrated on many levels almost every day. Frequent harvest rituals of "thanksgiving" were held throughout the year, because it was important to give thanks to the Creator and to the spirits of the foods taken—especially if one wanted to have more, and have these cycles continue. A failure to celebrate might bring on a period of drought and starvation.

The Cherokee Indians, descendants of ancient mound builders in the southeastern regions, have a traditional legend about the origins of medicine plants. They tell of a time when humans were increasing and hunting more animals. This caused the animals to become alarmed. They felt they were being pushed out of existence as humans took up more and more of their environments. Finally the animals decided to take revenge upon the people by inflicting them with many ailments, like arthritis, rheumatism, respiratory problems, cancer, and much more. The plants, however, overheard this scheme and thought this was harsh retaliation. They felt sympathy for the people and found a way to come to the rescue. Each plant offered some medicinal remedy. Earthmaker originally created plants with many valuable attributes. People simply needed to learn which plants healed each ailment and how to combine plants in formulas for healing remedies. Perhaps because of this, Indian People have always felt the need to celebrate the hunt and important harvests.

> Archaeological evidence shows that prehistoric people in the Great Lakes regions used an amazing number of their plants. They used 130 species for foods, 275 species for medicines, 18 species in beverages and seasonings, 27 species for smoking, 31 species as magical charms, 25 species for dyes, and 52 species for utilitarian needs. Another site in New York State yielded remains of 28 species of mammals, 5 species of fish, and 13 species of birds and turtles along with numerous plant species used for foods.

Early European travelers extolled the beauty and purity of American rivers filled with monstrous fish so large that they could threaten to capsize their canoes. They wrote about deer so plentiful that they roamed in great herds and wild turkeys traveled in great flocks. Hunters raved about the abundance of bear meat.

Darkening flocks of passenger pigeons (now extinct) were noted in 1810 to blanket a 40-acre forest of trees so densely that branches broke off under their weight when they settled to roost before evening. Hunters spoke of how the flocks darkened the sky when they flew overhead. America's inexhaustible resources must have seemed miraculous to many. They certainly sustained the mound builders richly.

The Great Temples and Expeditions

Centers of ceremonial life were the great central plazas and the great temples atop the huge temple mounds that faced the plaza. This was true in each site throughout the mound builders' world. These were independent city-states giving appreciation to each central ruler. No main axis of power governed everything, not even Cahokia, although Cahokia was the largest and most prolific of all temple mound sites. The various cities functioned as independent states. Perhaps it was their very autonomy that enabled them to endure and prosper as long as they did.

The mound builders dominated the one region where the Spanish were defeated. The Spaniards launched a series of expeditions between 1513 and 1574 into Florida, which was then the entire southeastern area of what would become the United States. Expeditions led by Juan Ponce de Leon (1513 and 1521), Panfilo de Narvaez (1528-1536), and Hernando de Soto (1539-1542) decimated many native people and left remarkable recordings of early observations. Spanish travel diaries are illuminating, as quotes throughout this book attest.

De Soto's expedition on the Mississippi River in 1542 encountered a fleet of 200 large dugout canoes, each filled with many men. The Spaniards were impressed with the numbers and skills of those warriors. The leader of the expedition sat on a raised platform under an awning at the rear of a large barge, giving orders to his force. Warriors stood with their bows and arrows ready. Their bodies were painted with red ochre and they wore great bunches of white plumes, while others sheltered the oarsmen with ornate feathered shields. The Spaniards remarked that the Indians appeared like a famous armada of galleys.

De Soto gravely misjudged these warriors while he was beating a hasty retreat down the Mississippi River to the Gulf of Mexico. He had offended their ruler of Quigaltam (perhaps a satellite of the Natchez) and his allies from a major mound building center in this region. They had sent these warriors to hamper de Soto's retreat. He sent Juan de Guzman with twenty-five men in armor to drive the warriors away. The Spaniards reported that:

...they took Juan de Guzman, and those who came ahead with him, in their midst, and with great fury closed hand to hand with them. Their canoes were larger than his, and many leaped into the water–some to support them, others to lay hold of the canoes of the Spaniards, to cause them to capsize, which was presently accomplished. The Christians falling into the water, and by weight of their armor, going to the bottom; or when one by swimming, or clinging to a canoe could sustain himself, they with paddles and clubs striking him on the head, would send him below.

De Soto's forces of 600 soldiers and 220 horses suffered heavy losses during these explorations, and he died of fever on the Mississippi in 1542. Only 300 of his men were able to retreat to Spanish territories in Mexico.

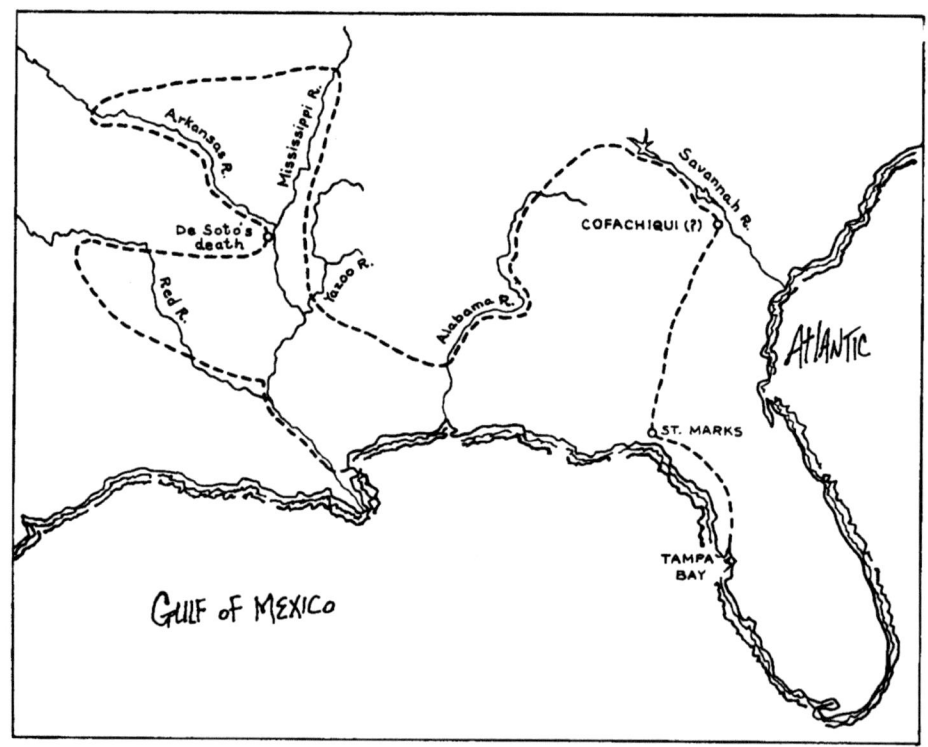

Map of Hernando de Soto's explorations (1539–1542).

History clearly shows that they, as well as other expeditions among the mound builders, ended in disaster for the Spanish. This must have seemed astonishing to the Spanish, who shortly after coming to the Western Hemisphere seized the centers of the Aztec, Mayan, and Inca realms, thus opening their opportunities for conquest in Mexico, Central America, and Peru.

The great temples of the mound builders' exalted centers never failed to awe the Spaniards, even as they hastened toward their own decline. Native lifeways seemed destined to move away from these vulnerable settlements with royalty and caste systems (four different social layers) into more egalitarian independent camps and settlements.

Early observers often called Indians "the Red Men," doubtless because the warriors and leaders they encountered were invariably painted with red ochre. Red was a sacred and exalted color throughout the Indian Americas. Warriors were the elite athletes in native societies and were often seen in beautiful attire, bodies scantily clad and well oiled and painted. Skirmishes and wars were great celebrations from which one hoped to return, but if not so lucky, then it would be a heroic way to die.

Puberty Rites—Coming of Age

Early Indian societies celebrated each important stage of life, as many of us still do. One of the most important rituals was the "Coming of Age" for young boys and girls as they approached puberty. Often the childhood name would be "thrown away" and a new name would be given to each youth at the dawning of puberty.

This was perceived as the dawning of their fertility, especially in earlier societies when marriage and child bearing began earlier and life expectancy was much lower than ours is today. It was not unusual to marry by the age of 14 or 15 and begin having children. In many societies infant mortality was high, and many Indian tribes understood the concepts of contraceptives and birth control practices.

> Great celebrations around **Native American puberty rites** are still performed today by the Apache, in the *GA'AN* Festivals, and the Navajo, in the *Kinaalda,* for their young women, as well as among many other tribes. The Apache Dance of the Mountain Gods inspires the following story because of its primal feeling of being carried forward out of antiquity into modern celebrations to honor young women in the tribe.

They Danced Me out of Childhood into Womanhood

> This is a fictional account of an 11-year old Cahokian girl, who is just "coming of age" (entering puberty). What was it like in ancient Indian communities? This is generally patterned after the Apache and Navajo Puberty Rituals as the most publicly celebrated rites in native communities. We can only imagine how the mound builders celebrated their young women during this most essential life passage.

I could feel my body changing. At first I did not understand why my breasts were growing larger, and sometimes cramps stopped me from running hard. My dreams began to change, and I had no appetite for my mother's delicious food. I often felt tired.

One day my aunt, my mother's sister, walked me down by the river. She talked about these changes and explained that we Turtle Clan women would gather soon to plan a big ritual for me and other girls my age. "You will begin to bleed soon. Do not be startled or upset. I will take you to the Moon Lodge," she said.

My mother also prepared me for these changes. "You will have increased powers and less energy on the first few days of your bleeding," she cautioned. "You must take good care of yourself at all times, and you should not touch the family's food while you are bleeding. That is your time of purification, or cleansing. You must not contaminate our food or preparations during these few days each month, nor work in the garden," she said firmly. I was stung by these ideas.

I went to the Turtle Clan Moon Lodge with my mother and aunt during their bleeding that month. This small, intimate lodge was separate and away from our houses. We had to stoop to enter it and sat comfortably around a small central firepit, The women shared their tales and reflections. We relaxed and talked together for two days, until the bleeding was finished. I almost looked forward to beginning my menses. Everyone knows now that I am becoming a woman. I am embarrassed and yet proud.

My father's Wolf Clan relatives began making plans for a huge ceremony at the summer solstice. He said, "We have chosen White Deer Woman, the Medicine Woman, to guide you through your puberty rites." I was thrilled! I had always admired her and wished I could be like her some day. Everyone respected her.

"The puberty rituals will last for seven days and each day you will be molded and prepared for your new life," he said. "We will gather much pollen from the corn and cattail

plants to use then. We must also go hunting for extra meat to feed the many people who will be our guests for your celebration. We need venison, fish, geese, and rabbit."

Soon my time of bleeding began. I felt self-conscious and had many mixed emotions, yet my mother and aunts supported me with stories. They taught me to eat wild mint leaves and chew on willow bark to relieve my cramps and headaches. They talked to me about the puberty rites to help me understand this enormous change in my life. I remembered attending these rituals last summer sponsored by the Crane Clan.

Just before the summer solstice my rituals began. Other girls my age joined me, each one supported by her family, clans, and a chosen medicine person. The entire tribe hosted us. The chief and his advisors made beautiful gifts—giving each girl a sun symbol ornament to wear. My mother's clan made me a glorious soft deerskin dress with fine beadwork, long fringes, and a small, beaded medicine bag.

Huge fires were stoked in the central plaza while tribal drummers beat a strong chant for the "fire dancers"—who mesmerized everyone with their rhythmic dances. Then at midnight the dancers came to draw each girl out into the plaza to dance with them. Each clan danced their support. We danced all night until the sunrise. Then we all retired to sleep for a few hours. Midday we gathered and shared a great meal together.

Each day these rituals were repeated and the energies heightened. I was dusted with pollen each morning and urged to run for miles toward the rising sun, and then return to rest and sleep. On the last day White Deer Woman danced me out of my childhood and into womanhood. She gave me the new name of Laughing Woman.

Rituals and beliefs surrounding the "coming of age"~the beginning of puberty~were no doubt vital in early societies. Marriage and childbearing began much earlier than in our modern society, and people's life expectancy was shorter than ours today. In ancient Cahokia, a young woman might have married (as we know it) by the age of 14; the young man too, or perhaps just a few years older. Infant mortality might have also been high.

Perhaps this young couple might have married at 14 and 17 and had three children before they were 23. If both parents survived, their old age would have been around 40. If one parent died, the other would perhaps look for another mate, simply because it was easier to have a partner in life than not to have one.

Life in mound builders' societies must have also been protected within family and clan associations. Both the mother's and the father's clans were important guiding forces for young people. Yet, political powers also interceded in young lives according to several later Spanish accounts that indicate the importance of marrying outside one's own culture group for status. This record of de Soto's expedition among mound builders in Arkansas territory in 1540 details how alliances were strengthened by marriage between a major leader and a woman from the allied ruler's lineage.

> *The Chief of Casqui came the next day, and after presenting many shawls, skins, and fish, he gave [de Soto] a daughter, saying that his greatest desire was to unite his blood with that of so great a lord as he was, begging that he would take her to wife...*
>
> *[De Soto] rested in Pacaha forty days, during which time the two caciques [chiefs] made him presents of fish, shawls, and skins, in great quantity, each striving to outdo the other in the magnitude of the gifts. At the time of his departure, the chief of Pacaha bestowed on him two of his sisters, telling him that they were tokens of love, for his remembrance, to be his wives. The name of one was Macanoche, that of the other Mochila. They were symmetrical, tall, and full: Macanoche bore a pleasant expression; in her manners and features appeared the lady; the other was robust.*

My First Vision Quest

> This is the fictional account of an 11-year old Cahokian boy, who is just "coming of age" (entering puberty) and looking at the ways it will change his life completely. Native boys were tested in many ways to help shape their character. More accounts of Native American boys' puberty rites exist, which were (and still are) rituals of endurance, learning personal strength, and seeking a vision of how one's life should be directed.

The hot summer wind pressed against my bare chest as I walked up the narrow trail from the river. My arms and legs have grown long and lean these last few months. I wonder how tall I will be? I'm almost as tall as my father now.

My dreams are changing, and in them my ancestors are coming to talk with me about personal things, changes, spirit guardians, courage, and new pathways...

I know the time of my fasting for a vision is coming soon. My uncles have been preparing me for this in various ways for many moons. Walks With Spirits, our Eel Clan Medicine Man, has picked a remote place on a bluff above the Mississippi River, where he will guide me in the beginning. Then I must continue alone. The two of us will build my first sweatlodge soon and I will begin to experience the periodic sweat-

lodge rites of purification and prayer. I must be strong and ready. I am already a good runner. Perhaps I will be a tribal runner like my uncle, who takes messages from our chief, The Great Sun, to some of the other villages far to the south of us.

I am fasting in preparation, going without food for a day or two, building my stamina. My endurance will be tested. I must not embarrass my family or myself. I will not fail. My coming of age rites will change me and take me from where I am now into adulthood. I am rapidly becoming a man. I look at my little brother and remember when I was younger and just played all day with my friends. I miss that now.

I will go to the vision pit above the Great River and cry for a vision. I will fast and pray to the Great Mystery for four, five, or six days—until the visions comes to me—to tell me what my pathway through life will be. I need to know!

> **The Vision Quest** is an amazing coming of age ritual for many American Indian boys (and some girls) today. It is usually personal and private. Every tribal group and clan has unique variations. The first vision quest occurs (usually) at the time of puberty; some believe that individuals might again seek to have a vision quest every 7 to 10 years, or at the time of critical passages in one's life that presage change.

Men and women made and wore beautiful jewelry made from the bones, antlers, pearls, stones, and shells from their natural environment, along with exotic trade items like copper, silver, and grizzly bear teeth. Similar items were also made for their children. This must have been, in part, to honor the spirits of these game animals, and to honor the earth spirits for all that was provided to make Cahokian life rich and successful. There was always the need to honor the Creator, by living and dressing well, in order to assure that more resources would always be available for Cahokian hunters, gatherers, farmers, and artists. Everything was geared toward appreciation and hopes for renewal.

Each season of ripening food was celebrated—from the first strawberries, to the marshgrass seeds, to the green corn, and then the gourds and pumpkins. Festivals with special songs and dances were enjoyed periodically throughout the growing season.

Special times in each year's passage were celebrated with "socials" and festivities that brought families, clans, and bands of mound builders together for communal meals, dancing, games, and shared rituals performed by the leaders. One of the greatest celebrations was for new life—when a child was born.

A Midwife's Tale

> This fictional account of a mature Cahokian woman, a midwife, who was an orphan brought up by her grandmother, is reflecting back on her life. She might be 50 years old. Long before there were doctors, women delivered their own babies, or were helped through the labors of childbirth by a midwife. This was usually a woman (rarely a man) skilled in the knowledge of "birthing" routines. These were (and are) highly respected people in every community. Midwives must have been very valuable among the mound builders, especially as people lived together in closer societies.

I have a deep love of new life, and as a child, the process of giving birth fascinated me. I found the matted grassy area at the edge of the woods where a doe gave birth to her new fawn. She did not see me lying in high grass nearby, motionless, watching the whole process. I would often return to this spot in spring to watch the birthing of fawns.

When I told my grandmother Nokomis about this, she asked if I would like to assist her at the next birth in our village. I was so honored to be invited. We walked together one night to the birthing hut of the Loon Clan where a young woman was just beginning labor to deliver her first baby. The Loon Clan birthing hut was located a short distance from the settlement of houses in a quiet grove of maple trees near the lake. It looked like a huge inverted basket made of bent saplings covered in rush mats and just big enough to hold perhaps six people comfortably.

Nokomis talked to the young woman softly explaining what was happening to her, and what a vital time this was in her life. She was lying on a pile of soft hemlock branches and moss, which smelled so fresh. She was having rapid contractions that kept her doubled up near the small campfire in the center of the hut. Her mother sat with her cradling her head, stroking her neck and back, and singing softly. I tried to help, too, and sang softly with her mother.

Nokomis began to massage her shoulders and neck with fish oil, wild mint, and yarrow from a small pot she had brought. As the young woman relaxed, Nokomis massaged her stomach and breasts with these same ingredients. This made her skin glisten and shine in the firelight, and enabled Nokomis to feel the baby's size and position. This continued for some time. The hours flew by, but we were not tired or sleepy.

Just before dawn Nokomis asked me to fetch the soft doeskin she had brought with her medicine bag. The young woman's contractions had quickened and she was moaning softly and trying to sing. Her water broke—spilling out into the birthing pit just beneath her, opposite the campfire. I was not surprised but impressed.

The two older women helped the young woman squat in a crouching position over the birthing pit, which was dug about two feet deep into the swept earthen floor and lined with fresh moss. After a few more hard contractions a tiny bundle of newborn energy slipped out into Nokomis's waiting hands. A baby girl! Nokomis immediately wrapped her in the soft doeskin, while she was squirming and crying. I watched this all with amazement!

Next they helped the new mother to stretch out on a bed of fresh evergreens, and handed the baby into her waiting arms and warm breasts. The baby girl began to suck at her mother's breast. Nokomis explained that this would help the contractions to quicken and expel the afterbirth sac, and also enable her bleeding to stop soon.

Nokomis got up and walked to the opening of the birthing hut and rolled back the woven reed mat. The coolness rushed in. Night was just ending. The last slip of moon was hanging next to the morning star. Sunrise would come within the hour and the village was awakening to this new day. Nokomis breathed deeply the cool crisp air and offered cornmeal and prayers to the Great Mystery for another successful new life in our village. Then she returned to us and offered more cornmeal blessings and prayers for strength and good health.

Then she addressed the newborn and said, "Thank you for coming to our village! We welcome you with prayers and love. We hope you will stay with us a long time and enjoy a beautiful life." I felt tears of joy! Nokomis later told me that this was how I, too, was welcomed at birth. Unfortunately my own dear mother died a few years later in a river flood. Nokomis and my aunts and father are my close family.

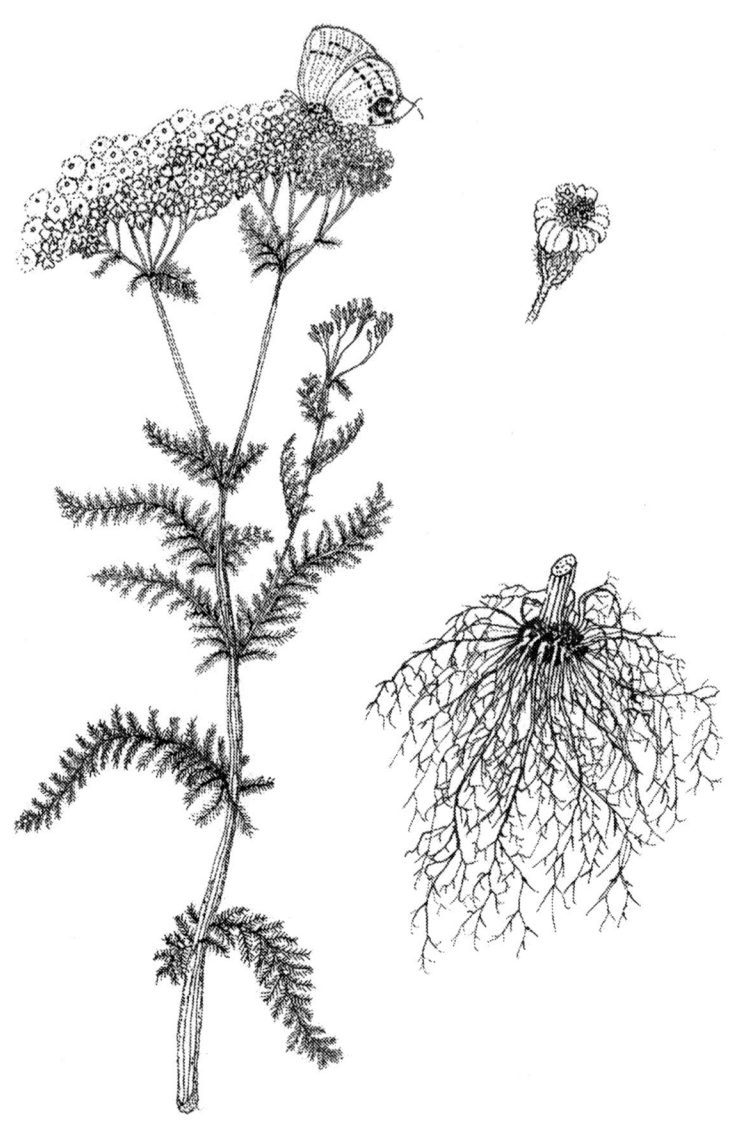

Yarrow, *Achillea Millefolium*, valuable perennial wild herb for calming & soothing the body; also relieves pain & acts like a mild anesthesia.

The young mother and grandmother asked Nokomis to please name their new baby. This was quite an honor and Nokomis was a respected elder. "Her name is Shenandoah," said Nokomis. "Her birth was attended by the morning star."

That was many years ago. I accompanied my dear Nokomis in many birthings before she passed into the Spirit World. She trained me to be a good, sensitive midwife. Now I carry on her legacy with great pride.

I have welcomed and prayed hundreds of new babies into our village and in the nearby city of Cahokia. I have lost only a few tiny lives, and that was always time for sadness. But sometimes a newborn comes into this world unwilling to stay, or its tiny spirit simply wants to slip away. It is a mystery. I believe that the ancestors' spirits take care of the lost babies' spirits in the Baby Hill inside the burial mound. They have given us special rites and songs to sing for them on the summer solstice.

~~~~~~~~~~~~~~~~~~~~~~~~~~~

Nokomis is Ojibwa for "grandmother."
Shenandoah is Shawnee for "Daughter of the Morning Star."

Celebrations of Death and Sacrifice

Perhaps nothing was quite so well celebrated as death among the mound builders in each phase and period of their amazing societies. After 4000 years of mound builders' existence, we know more about their death than their life, especially the deaths of their elite, through their burials. Considering the quantities and qualities of grave goods that accompanied elite burials, death was quite a celebration~and very big business!

Few ancient societies have ever left such memorials to their cultures and their honored dead. Some of the most elaborate burial mounds were those of the Hopewellian period. Honored dead were buried along with an astonishing array of grave goods. Finely finished objects and generous caches of raw materials accompanied elite burials. Scientists studying these sites believe that perhaps mound builders ceremonies were hinged in many ways to their extensive long distance trading network of valuable exotic materials. Copper and pearls were supremely important.

Mississippian societies clearly engaged in human sacrifice when a ruler or exalted noble died. There is no evidence that this practice occurred in earlier mound building burials. There was apparently more of this practiced at Cahokia than in other centers. Yet the practice obviously lingered among later mound builders, according to Spanish accounts. When de Soto died from disease near the Mississippi River in 1542, his men attempted to keep this a secret from local tribes. Yet the ruler knew well about his demise and visited the Spanish camp with two young men, whom he offered as sacrificial companions. He explained that *it was the usage of the country when any lord died, to kill some persons who should accompany and serve him on the way [to the afterlife]...*

You recall the poignant declarations of the Tattooed Serpent's wife in 1725. She willingly chose to accompany her dead, beloved husband and sacrifice herself to be buried along with him. This was in addition to his other two wives, his sister, his healer, his first warrior, his head servant and speaker (along with his wife), his nurse, and the specialist who made his war clubs. Many sacrificed themselves to die with him. This was perhaps one of the patterns of celebrating death when an important leader in mound building society died. Those people closest to you in life would accompany you into the afterlife, especially in major state burials when elaborate grave goods and ceremonies surrounded the passage.

When a ruler in temple mound society died, his temple (lodge or palace) was destroyed, and a new layer of earth was packed over the top of it all. The new ruler's temple was then constructed over this. The temple (or platform) mounds did not contain many burials and they were built up in many stages over many years. Most elite burials were made in conical or ridge-top mounds. This was an indication of how many ruling lineages and allies once inhabited each site and were aligned with the paramount ruler.

Death was sadness, but death, too, was celebrated! Loved ones were often buried with fine grave goods to support them in the Afterlife. Death was such an all-consuming aspect of life—the ultimate transformation—that it had to be honored. The separation of the deceased from those left behind often entailed ritual periods of mourning and preparation of the bodies, plus the fine objects to accompany the body along the road of death into the Afterlife.

[In] the regions watered by the Ohio and its tributaries…we find numberless mounds…. These are by far the most imposing class of our aboriginal remains and impress us most sensibly with the numbers and power of the people who built them…. The number of [mounds] may be safely estimated at 10,000. These observations were recorded by E.G. Squier and E.H. Davis in 1848, when they published their extensive survey of North American earthworks called *Ancient Monuments of the Mississippi Valley.*

Centuries later the Navajo (who eventually followed the mound builders and migrated to live in the Southwest) would see this as a transition of the spirit along the Great Star Belt (the Milky Way) to the Dance Hall of the Dead. The Path of Souls wants to be blessed with prayers, sacred cornmeal, and a small pot of corn soup, so that each departing spirit may make a happy journey. Perhaps these beliefs were very similar to those shared among the mound builders. We can only imagine, based upon their burial practices and fineness of grave goods, that the mound builders' spiritual beliefs about death and the afterlife were highly evolved and commanded great respect. Mound builders certainly had a fascination with the stars.

Sacred Societies

The sophisticated Cahokians must have had secret societies, aside from the healing, farming, and trading societies that men and women belonged to. Secrecy was often a way of protecting the personal power within an organization or medicine or healing formula. Dreams often foretold such societies and the mystical items of clothing, shields, musical instruments, tools or weapons that chosen members would make. A person could have a powerful dream in which he could see things he had never seen before, and get a sense of what to do about it. Dreams were an honored pathway of learning that native societies acted upon. It is quite likely that the mound builders held many of these same ideas.

Sacred societies were often secret groups where one could only be invited to join for very special reasons. Membership might require an endurance trial or tests of some kind, such as an endurance run for ten miles or swimming a long distance. Members might be tested by a long four or five-day fast, or spending a month (moon, or whole moon period) alone in a cave or out on a windswept bluff communicating with nature.

Reflections on the Hummingbird Society

> This is a fictional recollection on life by an older man, who was born outside of Cahokia, but welcomed into Cahokian society because of his special skills. He is perhaps 40 years old and reflecting back on the highlights of his life. His dreams led him to establish a secret society of "The Hummingbirds"—which carried out secret rites and rituals of healing to

> benefit the people. This is a glimpse into sacred ritual practices and the importance of secrecy in many Native American traditions. The account is loosely based upon Hopi and Zuni Kachina rites.

I used to be the swiftest runner in our little village beyond the evergreen bogs. This asset brought me to Cahokia as a young man, carrying messages from our village leader to the council of the Great Sun. Finally, after my tenth trip, they asked me to stay and become a member of their leadership council. Our meetings were always interesting, so I was honored to move to Cahokia and settle among the privileged class. In time I molded my life with the leader's beautiful daughter and we began our family together.

In those early days I started having powerful dreams and flashes of insight, which would frequently tell me what was going to happen several days before it actually came to pass. Whenever this occurred, there was always a hummingbird darting in and out—of my dreams or visions—as well as reality. I became more and more connected with hummingbirds, which are only with us here in Cahokia for half of each year. During the winter, when hummingbirds are gone, they would visit me in my dreams at night. They would show me and tell me things, amazing things about healing and life.

I talked with the medicine men about these things. They counseled me to start a secret society of Hummingbirds, whose members would honor the wisdom of the tiny hummingbirds and be fast in relaying their messages to the people. I even began to dream about how our regalia, medicine bags, and shields should look. My wife also had similar dreams and so the two of us began this small society and made our sacred items.

We occasionally brought new members into our group, depending upon their dreams and the messages that were shared. Most of the people in our surrounding city did not know about us, or our special work with dreams. Several members of the ruling elite joined with us. We built an underground circular room atop the Great Mound where we could all meet in sheltered privacy and work on our dream shields and rattles. This great room became our center, which we call a kiva.

We painted hummingbird symbols on our dance rattles with red ochre. We created small dance masks to wear during the big festivals along with special regalia. Everything concealed our human identities as we danced together in the main plaza during our spring rituals to welcome the hummingbirds back to Cahokia.

Sometimes it was difficult to carry this burden of secrecy. Not even our neighbors knew about our work, yet everyone was more secure because of our dreams. In time I realized that Cahokia had other sacred societies with distinctive rites, rituals and regalia.

~~~~~~~~~~~~~~~

> Hundreds of years later, Plains Indians would be famous for their Kit Fox Society for young warriors, and other unique men's and women's societies. These various social groups enabled the participants to visit closely together, make special ritual objects and attire, and bond more intensely with each other. This was similar to the close family bonds shared among the clans. All of this would be important if the villages were ever attacked by outside forces, or turned upon by unhappy trading partners.

Woodhenge

Around A.D. 1100 Cahokians constructed a circular sun calendar of large, evenly spaced cedar log posts, standing almost 20 feet tall, and painted red. Further archaeological investigations showed that four more "woodhenges" were constructed during a single century at Cahokia. The largest may have been 60 telephone pole-size posts set in a precise circle around a center post. Other "sun circles" may have included just 48 outer posts in a perfect circle around one center pole. Each post was dug deeply into the earth in order to stand up straight. Native artisans burned each end of the post in order to seal and protect it from decay. Fire was also used to fell these large trees and the charred wood was chopped away until the tree fell over. Some authorities call these "sun circles" and suggest that ancient Cahokians used this circle for calendar readings to plan vital annual events.

Posts were set to mark the primary cardinal directions of north, south, east, and west. Then additional posts were added to make a perfect standing circle around the center pole. It must have taken many strong men working in unison to erect these posts after the engineers and mathematicians determined the precise location for each.

As the sun, or moon, passed overhead in a clear sky, shadows would be cast upon the ground by the tall posts, which might have served almost like early hands on a prehistoric earth clock. This solar/lunar calendar/clock would have told observers, and perhaps a priest standing in the middle of the circle beside the center pole, the time of year based upon where the sun rose and set.

We have no idea what the ancients called the structure, which they used during certain ceremonial periods important to their agricultural way of life. Later observers called them woodhenges because of their apparent functional similarity to Stonehenge in England. Sky Watchers probably used these great circles to determine the changing seasons, especially the solstices and equinoxes, and sunrise and sunsets. Within the woodhenge Sky Watchers could chart the slow progress of each season and set planting and harvesting times.

The most important posts would mark the first days of summer and winter (the solstices), and posts halfway between these would mark the fist days of spring and fall (the equinoxes.) Perhaps the remaining posts marked Cahokian calendar points and moon positions throughout the normal year. There is so much we do not know.

Many early cultures had Sky Watchers who could foretell important events. Early observers called them priests because of their importance and elevated positions within each tribe. Woodhenges were impressive examples of prehistoric Indian science, math, and engineering skills that must have been used as solar/lunar calendars.

The Solstices and Equinoxes

Sunrises are most spectacular at the equinoxes when the sun comes up due east. Within Woodhenge at Cahokia one can see the sun rise precisely over the top of Monks Mound. Imagine an elite ruler standing atop this temple mound. He or she might have appeared to be giving birth to the sun a thousand years ago.

Mound builders honored the key marking points in the solar year. Ceremonies were planned for the first days of summer and winter (the solstices), as well as the first days of spring and fall (the equinoxes). Harvest times and planting times were foretold in these essential ways so that fertility rites and renewal rites honoring Mother Earth could insure the peoples' continuing success.

Mound builders also honored the sun, moon, stars, and special heavenly happenings, like comets, eclipses, and super novas whenever they occurred. Within each clan there were probably special women's societies, as there were also men's societies, and particular societies for young men and young women in order to help them prepare for life's many opportunities and knowledge of the cosmos, or heavens, was vital in many of these groups.

Remember you are walking on 'holy ground,'...we ask that you respect our culture and traditions.

—William Tallbull, Northern Cheyenne elder and religious leader

13

Creative Reflections~Legacies from the Ancient Mound Builders Worlds

Starting with the very first contact between Native people and non-Natives, we Native people have shared all we had—we have sacrificed more than our fair share. We have sacrificed our land, timber, water, and our mineral rights.

—John E. Echohawk, Attorney-at-Law, Executive Director
Native American Rights Fund, 1995

> There are amazing legacies left by the diverse mound builders and abundant mysteries. Why did they totally disappear as distinctive Native America societies after achieving so much? Where did they go and who are their modern relatives? We might never know the answers to some of these questions, but science has uncovered many secrets and continues to work on others.

Ceramics and carved stone figures of women recovered from Mississippian sites are exquisitely made and seem to suggest a special reverence for women. Many of the potters were quite possibly women who portrayed each other, or their daughters and themselves in classic roles. Some specialists speculate that ancient Mississippians may have developed a wide spread religious cult celebrating the "Earth Goddess" who was the bearer of life and symbol of all fertility. The famous Birger figurine (pictured in this book) portrays a woman kneeling on a circular base hoeing what appears to be the earth. Yet closer inspection reveals that the "earth" is a feline-serpent, a symbol of the underworld. This woman carries a squash on her back and a gourd vine encircles her shoulders. Was there a strong link between the powers of the earth goddess with their elite social rulers?

Other stylish objects appear to represent transformation, metamorphosis, and travel to different worlds. Shamanic symbols richly permeate mound builders' art in the form of effigy pipes, helmeted figurines, effigy bowls, pots, gorgets, and carved wooden masks and statues. Hammered copper effigies of key animals, like the hawk and buffalo fish, seem to signify connections with the Upperworld and the Underworld, which were and are primary shamanic realms for accessing knowledge from the unknown through entering states of non-ordinary reality. The ancient Mound Builders obviously had great facility in moving back and forth between the realms of consciousness to communicate with their gods, goddesses, and ancestors' spirits. The shamanic realms opened many more pathways to them for exploration and communication.

Pipes & Smoking

The mound builders created a great number of distinctive effigy smoking pipes that serve as ancient reminders of another way of life. Smoking was a ritual central to their ceremonial, religious, and shamanic practices. Effigy pipes were sacred vessels in which tobacco and other fragrant herbs were smoked. Tobacco was a sacred plant and it is still considered to be so by most traditional American Indians today. The wild varieties are the most sought after by traditional folks for ceremonial uses, which included more than fifty species and varieties of *Nicotiana,* the Latin genus name of tobacco.

The pipe embodied clan symbols, special visions, dream creations, shamanic powers, and personal charms. The talented mound builders carved the creatures from their universe with great skill and ingenuity; then held in their hands the power of these 'spirit beings' as they held counsel and contemplated their life's work. The finely carved stone effigy pipes represent some of the best Mississippian artistry.

The mound builders made an exceptional number and range of pipes from clay and stone. Some of the clay elbow pipes (bowl at right angles to the stem) were probably made for smoking in private or in tribal settings, and the smaller ones intended for personal enjoyment. The Mississippians had a strong tradition of pottery making and wonderful examples of these pipes were most numerous in Missouri and Arkansas regions. The most artistic examples are the carved stone pipes fashioned from select 'soft' stone that does not crack or shatter in extreme heat. These varieties include limestone, steatite, pipestone (Catlinite), and Bauxite. The famous Indian painter George Catlin observed the pipe making process in the early 1830s among central Plains Indians. Later the red pipestone was also named Catlinite in his memory.

> *The Indians shape out the bowls of these pipes from solid stone, which is not quite as hard as marble, with nothing but a knife. The stone is of a cherry red, admits of a beautiful polish, and the Indian makes the hole in the bowl of the pipe by drilling into it a hard stick, shaped to the desired size, with a quantity of sharp sand and water kept constantly in the hole, subjecting him therefore to a great labour and the necessity of much patience.*
>
> —George Catlin, 1832-39

Pipes were supremely important because so many of them are imbued with great technical skill and craftsmanship. Hopewell and Mississippian pipes are especially highly embellished. Perhaps by looking closely at the great range of pipes we can learn more about their religious thoughts, science, and technology. Some of the human and animal effigy pipes seem to suggest man shapeshifting into the animal, and then back again into his own personal form. This is an important aspect of shamanic work.

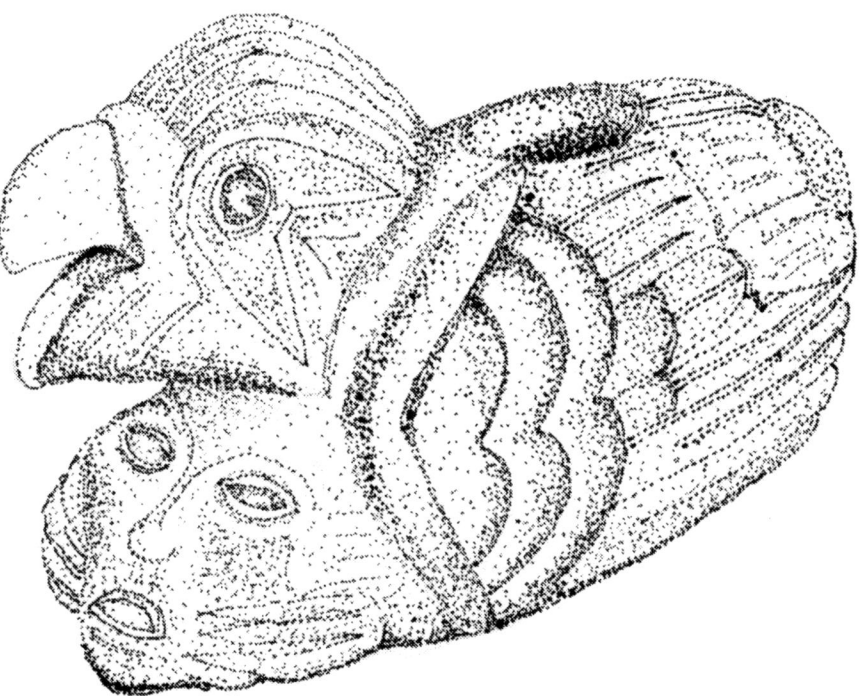

The Raptor and Wo/man Effigy Pipe, 6.25-inches long; famous Southern Death Cult pipe carved of limestone; note falcon's weeping eye motif; found in Mississippi. Symbolizing the power of a shaman shapeshifting. (Thomas Gilcrease Museum, Tulsa, OK)

Wild Tobacco, *Nicotiana rustica* **& Wild Coyote Tobacco,**
N. attenuate
Valuable ceremonial & medicinal plants, & desirable trade items.

Tobacco and other smoking herbs were consumed in a variety of ways and for many different purposes. Tobacco was primarily an offering substance to the Creator and to the supernatural world. Its fragrant smoke while burning would rise up into the sky carrying one's thoughts and prayers. It was a bridge between man and the supernatural world, wherein appeals and associations were made and established for the benefit of the family, clan, and tribe. Smoking was obviously done for pleasure, and for ceremonial, ritual, and health reasons~alone, or with a group of people. The communal smoking of tobacco created a bond within the group, especially for hunting or ritual planning. Tobacco was a gift between people as well as an important trade commodity.

The presence of so many stylistic smoking pipes seems to suggest that shamans and leaders might have smoked narcotic forms of wild tobacco, and possibly other mind-altering substances. Various species of wild tobacco populate southern regions where Aztec and Mayan cosmologies each celebrated "Tobacco Gods" who were involved in healing rituals and shamanic rites.

Kinnikinnik is an early Cree Indian word for "mixture" of botanical ingredients. This included many fragrant, dried herbal substances like scraped willow, dogwood, and sumac bark, bearberry leaves, angelica leaves, blueberry leaves, and yarrow leaves and blossoms, along with aster and goldenrod leaves and blossoms, bee balm, lobelia, wild lettuce, sunchoke and sunflower leaves, and mallow leaves. Some of these mixtures would include up to 30 or more different botanicals, each dried and shredded, then blended and carried in a special smoking pouch. Kinnikinnik may or may not have included some tobacco leaves. In some regions where bearberry was common, this plant alone was called kinnikinnik because it was so highly prized. Kinnikinnik was carried and used for ceremonial purposes, and it was also smoked in pipes. Perhaps kinnikinnik, like many other Indian words, originated among the early mound builders.

Bearberry, *Arctostaphylos uva-ursi*—an important ceremonial, smoking plant, & valued medicine, & also an important trade item.

Legacies in Earth, Bone, Stone, and Pearls

While Cahokia declined and was abandoned by the early 1400s, other temple mound cities continued to flourish well into the 1600s. Spanish accounts in the early 16th century recorded fascinating encounters with several Great Suns presiding over remarkable temple mound cities and villages south and east of Cahokia.

Hernando de Soto explored an abandoned temple mound center, in what would become South Carolina, in about 1540. It was perhaps a tenth of Cahokia's size, and had been unfortunately decimated by a disease epidemic before de Soto found it. The Spanish soldiers explored the empty village, including the great temple lodge atop the temple mound. They recorded that it was a long, single story building with a high ceiling. The roof was covered with conch shells on fine cane mats. Two rows of carved wooden warriors stood just inside the large doors, and they each held remarkable weapons, from battle axes and broad swords, (all carved in wood) to clubs, pikes, bows and arrows, maces, and batons. The Spanish soldiers also found valuable bundles of furs and chests filled with freshwater pearls. Searching further they found eight annexes filled with large stores of shields and weapons.

> Around the world, among all the people who used **pearls** in adornment and ceremony, no other group used as many as the ancient Hopewellians and Mississippians, who made quite remarkable industries of their many species of freshwater mussels. Pearls were drilled and worn abundantly as beads by elite men and women. Pearls were also used to adorn ceremonial objects, such as the eyes of birds in stone and clay effigy pipes, and fastened into other objects.

The ancient mound builders have left our own young civilization many gifts to ponder. Perhaps the mounds and the ancient mound builders have given us a deeper sense of nature and the sacredness of earth. Their cultures continue to inspire our modern society in many ways—from the spiritual connections to the mounds and creating sacred space, to the wealth of material objects created from their visions and dreams.

These cosmopolitan farmers, hunters, artists, and warriors created great wealth in an economy devoid of money and without the wheel. They thrived on far-flung trading networks. Mound builders societies prospered through systems of barter and exchange, in which people were celebrated for their natural talents and creativities. Men were generally the leaders and warriors, and women were the "power brokers behind the thrones." Men and women were probably very equal in mound builders societies. They were vital partners in each family's and community's survival.

Native People told many stories, which served to explain the seemingly unexplainable things in life and nature. Some were "creation" stories, which explained how the world was created and how things became the way they are now. Stories have always been teaching tools serving to educate listeners about many things. The following story comes from the rivers and lakes of eastern North America, and may have been carried forward through time from the ancient Mound Builders.

The Lilies of the Lake

> This old Woodland story describes how some stars chose to live on earth and become water lilies. It is charming to see how native people wanted to interact with the stars and the star world, much like in the ancient Greek and Roman times. Stories like this served to build personal bridges between Native People, their heavens and their environments. Many native stories tell about the origins of things, especially stars, the moon and sun. The roots of water lilies were an important source of food for Indian people, especially during winter and spring. And, Plains Indians believed that puffballs were actually stars that had fallen to earth to benefit people's needs for foods and medicines.

Far back in the earliest times, there was a beautiful Indian village beside a fine lake. The people were all very happy there. Each evening they would have dancing and drumming around a large central fire. Then the storytellers would spin fine tales and everyone would gather round and listen.

Well, these stories and rituals became so fascinating that even the animals would slither and crawl in near this village to listen, because they enjoyed native stories. They are often featured in these stories.

Some say it was the drumming and then the stories that first attracted many of the stars to drop down closer to this village and hover over the lake listening. Some say that the stars also became fascinated by their own reflections in the calm lake water, and lingered too long on warm summer nights. Every now and then a shooting star would arch out over the lake and plunge into the lake.

Soon the leader of the Sky World became alarmed that more and more of his stars were slipping away to Mother Earth at night, so he cautioned them not to do this. He asked Grandmother Moon to talk to them, too. She was very kind and reminded them of their roles of lighting the night skies and helping the ancestors spirits find the great bridge into the Afterlife.

The stars were grateful and behaved for a while, but then they found they couldn't resist the pull of the drums and the lure of native stories, and the allure of their reflections in the lake waters—so they began slipping away again.

Grandfather Sun became concerned and told them that they must choose whether they would remain stars shining in the night, or leave the Sky World and go settle on Mother Earth. He found it distressing to see them go back and forth across such distances. Many of the stars felt it was distressing, too. So quite a few of them took this liberty to leave—on good terms—and they sparkled down into the water to float on the lake near this magical Indian village.

The Creator turned them into water lilies, whose great white flowers shine like stars during the summer when they bloom. Through the centuries they have multiplied, just like the people whom they came to earth to serve.

The Mound Builders were ingenious and used abundant natural resources to their creative advantage. Many of their finely crafted objects have lasted for thousands of years and are in some of our finest museums and private collections. Many more items have perished and been reclaimed by the earth. Some sacred American Indian objects were created to be placed on earthen altars and be reclaimed by the elements as an honoring for Mother Earth. Certain masks, prayer feathers, and other sacred items were (and are) meant to serve a special use and then be reclaimed by Mother Earth (returned to nature.)

Earlier Time

The mound builders world was obviously governed by very different calendar concepts from ours. Perhaps they marked the passage of time from one full moon to the next, and from one summer solstice to the next. The most prominent environmental resource might have marked each phase of the moon. Traditional Native American calendars mark the 28 days from new moon to new moon, or full moon to full moon; and there are thirteen moons in a whole year. Eastern Woodland Indians say we are all living on Turtle Island, and there are thirteen moons to match the thirteen plates on turtle's back. Could this reasoning have come from early mound builders concepts?

Cahokia's calendar year might have felt a little like this:

(September)	Moon of Green Corn Harvest & to Gather Sweetgrass
(October)	Moon of Falling Leaves for Gathering Wild Mushrooms
(November)	Moon to Store Food in Caches for Thanksgiving
(Indian Summer)	Harvest Moon to Dig Medicine Roots
(December)	Old Moon When Trees Crack & Frost Spirits Roam
(January)	Cold Meal Moon (Natchez) & Ice Covers the Water
(February)	Snow Moon When Geese Return and Maple Sap Runs Sweet
(March)	Little Frogs Moon When Buffalo Calve
(April)	Planting Moon When Geese Lay Their Eggs
(May)	Strawberry Moon of Corn Planting
(June)	Moon of Summer Blossoms When Corn is Hilled Up
(July)	Moon of Ripe Cherries When Corn Is In Tassel
(August)	Corn Silk Moon When Beans and Squash Ripen

Each moon was vital for the period of vital natural resources usually available during its light. The yearly cycles of sun, moon, stars, planets, and earth's seasonal moods were intensely interesting to our ancestors. Close attention was given to natural events, into which periodic festivals, celebrations, tribal rituals, and normal lifeways were set. Everything in life was keyed to the rhythms of the earth. People would have marked their lives from one winter or summer to the next, and from equinox to equinox. Special ceremonies were set to honor these logical turning points.

This special song comes from ancient Seneca women's traditions of honoring Grandmother Moon, and was probably sung during the times of each full moon. This might also have been sung during the brief two days each month of women's "moon lodge"~when her menses caused her to withdraw from regular family life. This may have even passed down in words and use from ancient mound builders' origins.

Neesa, Neesa, Neesa,   "Grandmother Moon" (repeated, and sung softly)
Neesa, Neesa, Neesa,
Neesa, Neesa, Neesa,
Guy Whey Oh,   "Here we greet and sing her praises"
Guy Whey Oh.

[This comes from Seneca Medicine Woman, Twylah Nitsch.]

did they hold the vision?

Ghosts of their languages echo in names of
Rivers, states, places all over this land. Their
Stone blades, hammers, drills, spearpoints fill
Museums, historical societies, private collections.
Quartz and flint arrowheads still appear in cornfields.
Pottery shards peer out of soybean fields
Caught in primal silt flowing from rainstorms.

I offer tobacco and cornmeal blessings from
My heart and spirit; my soul sings compassion
To this ancient dirt, older than time, holding memory,
Feeding what is left of archaic bones and grave goods
Within this Indian Mound, still exuding palpable
Spirit Energy across a mystical millennium…

Did they hold the vision that this earthwork
Would stand memorial, consecrating countless
Generations following their early societies?
Did they see how great burial mounds could
Change the land and energies and remembering?
Their architecture and engineering humble
Observers dazzled by the ancient sophistication.

Shaking the gourd, I chant another prayer
Softly into the haze becoming darkness, obliterating
All shapes and forms except for the vision I hold of
Lizard Mound dressed in new snow rising along
Wisconsin's windswept high prairie among
Many other effigy mounds and earthworks holding
Ancestors' spiritual concepts aloft in
Giant talismans on the breast of Mother Earth.

14

Conclusions~How Did These Magnificent Cultures End?

From time immemorial, Indian tribal Holy Men have gone into the high places, lakes, and isolated sanctuaries to pray, receive guidance from the Spirits, and train younger people in the ceremonies that constitute the spiritual life of the tribal community. In these ceremonies, medicine men represented the whole web of cosmic life in the continuing search for balance and harmony…

—Vine Deloria, Jr., Standing Rock Sioux historian, author, teacher

> The Adena People gradually evolved into the Hopewell Culture and some evolved into Effigy Mound Cults, while others became the Mississippians/the Temple Mound Builders. What caused the demise of these amazing prehistoric scientists, engineers, farmers, and artists? In spite of all their remarkable landforms and distinctive worldviews, these magnificent cultures abandoned their major centers and apparently melted away…and evolved into other tribal groups.

The End of the Mound Builders?

Despite the splendor of ceremonial rituals and the high quality of life, Cahokia's population declined and moved away. Eventually the City of the Sun lay in ruins—a ghostly metropolis—but a living monument to those who created it and once lived there. A land haunted by sacred earthworks lay open to the ways of nature.

Perhaps a combination of many things that our modern society can easily relate to caused Cahokia's decline. The Cahokians may have suffered from a series of disease epidemics, mineral and vitamin deficiencies, birth defects, or other abnormalities. A swing to higher infant mortality rates and childhood deaths might have alarmed people. Perhaps as cities do, Cahokia became vulnerable to her own garbage and human wastes that can breed diseases and cause unhealthy pollution problems. Farmers and hunters can make a life for themselves almost anywhere, rather than be crushed beneath a decaying city struggling to survive. Cahokians must have used up too many of their dependable natural resources.

Trade dynamics might have completely changed, leaving Cahokia no longer valuable in the far ranging systems of lucrative exchanges. Other trading centers may have taken away or diminished Cahokia's economic influence. Such pressures may have siphoned off vital parts of Cahokia's population.

Changing weather patterns brought drier, colder, droughtlike conditions to the region for more than 20 years. Internal strife, political turmoil, and warring political factions may have further eroded the city's strength. Imagine the toll exacted on Cahokia's leaders, medicine people, and shamans when they were unable to satisfy the people's basic needs and correct fickle turns of weather and fate. This must have shaken Cahokia's core and changed her political organizations and society structure.

Conclusions~How Did These Magnificent Cultures End? 197

Mississippian Deer Shaman's Mask from Key Marco, Florida
Carved from cedar wood, this mask combines human & deer features.

Mississippian life in the valley and floodplains of one of the world's mightiest rivers was vulnerable. Natural disasters, such as great floods, must have played catastrophic roles, ravaging Cahokia's outlying homes, economy, and social life.

Think about our own vulnerabilities in the modern floods of 1850, 1873, 1927, and 1993. Floods of such magnitude would have overwhelmed grain storage pits in villages, destroyed fields of vital crops, disrupted reliable harvesting of aquatic resources, and exacted a staggering death toll.

It also appears that this large thriving center simply used up too many of its natural resources, firewood, fertile garden soils, mussel beds, salt deposits, and clean drinking water—making it necessary for the remaining population to move away and settle in other regions, never to return. Yet we return to Cahokia again and again to ponder her greatness and her fate.

Cahokia dominated an area almost the size of New York State at the height of her power in 1150. This greatest of Mississippian cities still draws us like a magnet. Cahokia has been studied for over 200 years, and extensively, scientifically studied since the 1920s. And there is so much we still want to know about her and her people. Ancient Cahokia spread out along the Mississippi was in many ways similar to ancient Egypt spread out along the Nile. Each center of antiquity, trading, and art has blessed us with countless treasures and yet was cloaked in many mysteries.

Cahokia's Decline and Demise

For all her impressive achievements, this magnificent city could not survive. Over the centuries Cahokia depleted her once abundant natural resources. Drier than normal years were part of the changing climate, and this must have seriously challenged farming productivity, along with depleted soil fertility. Some environmentalists suggest that the forests were giving way to grasslands. Certainly Cahokians had harvested a great deal of their native forests for building and fuel. The great floodplain of the American Bottom had changed and could no longer support the density of life once so easy there.

Cahokia's City of the Sun lay abandoned and in ruins by about 1500. Only her amazing mounds remained to tell a partial story of her former greatness. Earthquakes or other natural disasters may have also shaken Cahokia's confidence.

No one knows where the dynamic Mississippians went after they left Cahokia. Many folks may have been absorbed into other Temple Mound societies for a while, yet even Cahokia's satellites and distant trading partners would experience similar social and environmental pressures. Yet mound building continued…

> **Cahokia Mounds** was designated a World Heritage Site by the United Nations Educational and Scientific Organization (UNESCO) in 1982, because of its importance to our understanding of the prehistory of North America. The site features a variety of special events, craft classes, lecture series, tours and other programs year round.

> Geophysicists, who gathered at the American Geophysical Union in 2001, suggested that **earthquakes** might have been a major factor in many incidents of cultural collapse around the world, or at least a trigger that pushed some major settlements over the edge. This would be especially likely if the ruling elite was weak or killed. Monumental structures get destroyed rather than the farmers and hunters, who go on.
>
> This theory seems particularly true for the Mayan People in Central America late in the ninth century when an earthquake may have caused the abandonment of two major Mayan cities: Quirgua and Benque Viejo. Certainly people living close to Mother Earth might read natural disasters as signs of termination with a particular place.

Beyond Cahokia, other dynamic mound builders' centers gradually experienced similar declines in population and the demise of a once grand way of life. The migration and evolution of the mound builders are mysterious points to ponder. Mound building did not perish with the collapse of the major centers. Perhaps the mound builders continue to evolve through the bloodlines of traditional American Indians today, who like the HoChunk in Wisconsin continue to bury their dead in modest mounds of earth, thereby creating sacred earthen symbols to their rising spirits.

The Southern Cult

Sometime around 1540 in response to de Soto's disastrous march across the southeast, a Messianic movement sprang up in retaliation to the Spaniards' "policy of calculated murder" and rape of Indian towns and villages. Others trace the cult to the shock waves after the fall of Mexico to Cortez in 1519, as many indications show a close trading relationship between the mound builders and people in Mexico. Scholars suggest the Southern Cult was reinvigorated and "stimulated by a culture crisis," said Kenneth Gordon Orr in 1946,..."developing from a knowledge and fear of aggression from European invaders." The Southern Cult had previously flourished at major centers across the southeast from about 1150 to 1350, and revived in response to major outside influences.

Other scholars who have studied the Southern Cult Religion suggest that the striking symbolism was an expression of vitality and vigor, not terror, and many designs are viewed as emblems of harvest, fertility, and renewal. Some rites foreshadowed the busk, the traditional summer green corn festival of the Creek Indians, descendants of the mound builders.

Almost 400 years later, scientists, dazzled by the bizarre and haunting mound metropolises with their amazing cult objects, would call them the Southern Cult, also the Southern Death Cult, or the Buzzard Cult, and the Southeastern Ceremonial Complex. All of these names apply to the series of sophisticated settlements, and their incredible handmade objects, across the south from Etowah and Ocmulgee in Georgia, and Moundville in Alabama, to Spiro in Oklahoma.

> *There is a touch of fantasy about the Southern Cult material that excites wonder and arouses a sense of widening vistas. The mounds become the platforms for lost cathedrals; the village folk are something more than diligent farmers to us, as we contemplate the regalia and paraphernalia of their religion. Their poetry, their history, the names of their kings, are forever lost to us; but here are wands, scepters, holy plagues, ritual vessels, all speaking of elaborate ceremonies shrouded by time, and all decorated with the nightmarish figures that these folk held sacred.*
>
> —Robert Silverberg

In Reflection

The ancient mound builders continue to cast long shadows across American landscapes and memories, and tantalize our imaginations and sciences with remarkable materials. This is quite a legacy from the diverse "stone age" people who developed sophisticated technologies, solar calendars, and engineered mammoth mounds to honor their leaders and loved ones. We look carefully into the faces of the anthropomorphic pipes, pots, and stone statues created by the mound builders to glimpse the likenesses of their diverse faces and forms. Were these items of pleasure or made for sacred reflection, or both?

The Mound Builders grow evermore appealing with time. Almost half the states in America still have ancient mounds. This earlier civilization evolved and flourished across North America's heartlands for more than 4,000 years. And their spiritual presence continues to haunt and inform our very American prehistory with dignity and awe. Their descendants live among us today as their bloodlines run through many American Indian tribal peoples.

The diverse Mound Builders exerted impressive environmental impacts across their ancestral homelands. Perhaps because of their creative spiritual earthworks, we are a more sensitive people today, aware that Mother Earth always has much more to teach us.

> *Smallpox, typhoid, bubonic plague (in 1617, 1633 and subsequently), influenza, mumps, measles, whooping cough—all rained down on the Americas in the century after Columbus. (Cholera, malaria, and scarlet fever came later.) Having little experience with epidemic diseases, Indians had no knowledge of how to combat them.*
> —Charles C. Mann, *1491,* in *The Atlantic Monthly,* March 2002

Following the period of "initial contact" with Viking explorers over a thousand years earlier was Christopher Columbus's claim of the New World in 1492, and John Cabot's claim of Newfoundland in 1497. Europeans quickly began visiting, exploring, and fishing the waters of the New World in great numbers. More than 30 French ships a year visited these waters by 1550, and by 1578 more than 300 ships arrived. Contacts increased and by the 17th century, colonies began sprouting all along the Atlantic coast: Cape Sable Island in 1598, Maine in 1604, Virginia in 1607, Massachusetts in 1620 and New Amsterdam (New York City) in 1624. Hostilities erupted, inevitably, between the cultures, plus a plethora of European diseases arrived with the fishermen and colonists effectively depleting Native American domination of the land.

Environmental Reflections—The Importance of Animals

The mound builders hunted and trapped, fished and gathered the abundant wild resources in each of their diverse environments. Their populations increased proportionally with the increase of healthy animal populations, even when farming became a major means of support. Later societies and groups were also dependent upon these resources

One French fur trapper brought in the 3094 animal skins recorded here in the summer of 1814. Imagine this being repeated on a large scale year round, year after year, by various trappers, Indians and 'mountain men' and regional homesteaders struggling to survive. We think of the fur trade as being largely centered upon the pelts of beaver, deer, mink and muskrat in the east, and hides of buffalo, elk, antelope, and bear in the west. Yet it was so much more extensive.

European traders and entrepreneurs opened the Mississippi River wilderness by establishing trading posts near government forts and fur "factories." Millions of dollars in luxurious furs were shipped along the great waterway.

A French Fur Trapper's List

The following list of furs traded by a French trapper Jac Porlier in July 4, 1814, recorded as peltries sold, noted in the Wisconsin Historical Society Collections; this gives some sense of the wildlife populations at the time. A *Livre* equaled 20 *Sols* and was worth about .20 cents

		Price	Livre
838	Chats—Raccoons	50	2095
47	Loutres—Otters	22	1034
77	Visons—Minks	50	192
5	Renard Rouge—Red Foxes	2	10
6	Pichous—Bobcats	2	12
29	Oursons—Bear Cubs	5	145
18	Peau Ours—Bear Skins	24	432
5	Peau Ours—Bear Skins	15	75
1670	Rats—Muskrats	30	2505
104	Liv. Castor—Lbs. Of Beaver	20	2092
29	Piguant—Fishers	6	174

123	Martes—Martens	4	492
2	Loup Cervier—Lynx	3	6
1	Renard Virginie—Virginia Gray Fox	2	2
74	Peau Chevreual—Deerskins	5	370
(66	Petits Rats Invendeus—66 small Muskrats were unsold)		

The total should be 9636 livres and 10 sols for 3028 pelts sold.

Rumors circulated, during the late 1840s, that the Chippewa living in the Lake Superior region of northern Wisconsin were destined to be removed and sent westward into inland Minnesota. A tribal delegation traveled to Washington, D.C. in 1849 to petition Congress and President James K. Polk for a permanent home in Wisconsin for the Lake Superior Chippewa. The delegates carried this symbolic petition with them on their journey. The animal figures represent various "totems," as determined by each family's lineage, each connected to the others' mind and heart to represent the unity of their purpose. These figures are also linked with the land and water of the tribe's beloved north woods. The blue waterline across the bottom of the pictograph symbolizes Lake Superior: the circular images at the lower left represent a chain of their wild rice lakes. This petition states that clan leaders, each represented by a drawing of that clans' animal symbol or lineage, are of one mind and one heart, (indicated by the lines connecting each of their hearts and minds), in wishing to not be removed from their wild rice beds in northern Wisconsin.

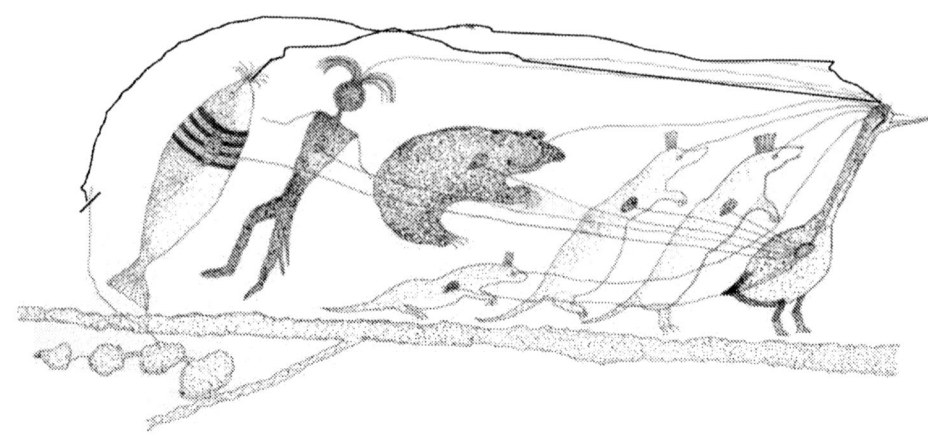

The Unity of Their Purpose—The Lake Superior Chippewa
This illustration was inspired from a drawing by Seth Eastman from Henry Rowe Schoolcraft's *Historical and Statistical Information Respecting the History, Condition, and Prospects of the Indian Tribes of the United States*, Vol. 1 (1851). The State Historical Society of Wisconsin, Madison, Wisconsin.

Portage la Prairie

West of Winnipeg at *Portage la Prairie*[1] a
cold wind blows the ancestors' spirits free to
a place where Assiniboine Indians carried
canoes across open land going to
English posts along Hudson Bay.
Hunters, trappers, fur traders crossed from the
Assiniboine River to Lake Manitoba in the
early 1700s along "Prairie Portage"~an
avenue to adventure and prosperity…
Portage la Prairie, Manitoba, is
homeland of the Saulteaux Ojibwa on Long Plain Reserve,
carving pathways beyond memory. Hear the loons call…
North America is dotted with *Portage la Prairies*
and other portage places skirting treacherous
rapids and impassable conditions along the
original highways, wild river systems.
Listen to spirit voices sing of countless
journeys on the Ohio, Wisconsin, Illinois, and
Tennessee Rivers—each carrying their special
Indian names into history. Come run the rapids along
the Alabama, Rio Grande, Colorado, Columbia,
Mississippi, and Missouri Rivers…see ancient
Spirits in the waves calling a traveling song.

---

1. *Portage la Prairie* is French for "Prairie Portage." As mentioned in Verendrye's journal, in the early 1700s it was a "place where the Assiniboine Indians carried their canoes on their way to the English posts on Hudson Bay." Situated west of Winnipeg, this portage was used by fur traders to cross from the Assiniboine River to Lake Manitoba.

   The Village of Whirling Thunder was a Winnebago summer village that once sat along the north bank of the river in Portage, Wisconsin. The Winnebago controlled the portage from here, selling rights to transport boats across it from the 1790s. Descendants of the Winnebago maintained a presence here and would move off to the blueberry plains in season, and to the cranberry marshes, and to hunt and trap.

Portage, Wisconsin,
windswept prairies full of snow and shadow
blanket awesome Indian effigy mounds.
You cannot touch the chiaroscuro
elements in flight on this land, or
the exiguous understanding
of it all…
breathe it,
paint it,
catch the mystery!

A Story of the Pleiades
(From the Eastern Woodlands)

> This early Woodland Indian story is about the origin of an important constellation. Many American Indian tribes see the prominent stars as spirits of important ancestors. Their traditions tell about star formations, which were often early super—humans who climbed into the Sky World to become prominent stars. The ancient Cahokians were noted for their "Sky Watchers"—priests who watched the heavens to learn more about the passage of sun, moon, stars, comets, and important constellations.

Back near the beginning of time, when the world was new, there was a fine Indian village beside the broad Mississippi River. The people there had plenty of game and wild mushrooms and plants to eat. They had a good life. The village was busy with women tanning hides, men making tools and spearpoints, elders sewing fine moccasins, and children running and playing.

A group of eight young boys, brothers and cousins, were very close and played together each day. As they grew, they started their own dance group and made their own drums and rattles. They made a special clearing in the woods near their village where they would gather to dance and drum. In time they became quite excellent. Folks could hear them drumming back in the village and appreciated their music.

One day the boys decided to hold a long festival and campout for several days by themselves. They asked their parents for extra food to take to the clearing in the woods.

But it was late in the season; hunting had not been successful, and there was little to spare. Their requests were refused.

The boys were not troubled. They continued their plans to meet and dance. Their music grew so mystical several nights later that their parents went out to find them.

The boys had been fasting, going without food and water for days. Their music and dancing became ever more intense. They grew lighter, and as they did their circle dance they began to rise up off the ground. They were intoxicated with their dancing!

By the time their parents and the other villagers arrived in the dance clearing, the eight boys had risen well up into the night sky—dancing and singing. They glowed with energy! The parents called out to them frantically. The last little boy turned to listen, and he became a shooting star.

The seven remaining boys could not hear the villagers calling to them and continued dancing up into the Sky World, where they became The Seven Dancing Stars, or the Pleiades, as we know them today. Native people watch for them as they circle the heavens, and know when they appear overhead in winter it signals the time of storytelling and remembering.

Stories reflect back to the ancient Mound Builders. Star legends and origin tales spin visual mind pictures of another universe. Listen to the stories and imagine life a thousand to four thousand years ago.

Some Indian Place Names in North America

Adirondacks	(Iroquois)	"bark eaters"
Alabama	(Choctaw)	"plant gatherers"
Alaska	(Aleut)	"mainland"
Appomattox	(tribal & village name)	"tobacco plant country"
Arizona	(Pima/Papago)	"place of the small spring"
Chattanooga	(Creek)	"rock rising to a point"
Chesapeake	(Algonquian)	"big bay"
Chillicothe	(Shawnee)	"village"

Connecticut	(Mohican)	"long tidal river"
Dakota	(tribal name)	"the people" a.k.a. Sioux
Idaho	(Shoshone)	"sun on the mountains"
Illinois	(tribal name)	"the people"
Iowa	(tribal name)	"the people"
Kalamazoo	(Algonquian)	"smoky"
Kankakee	(Mohegan)	"wolf land"
Kansas	(tribal name)	"people of the south wind"
Kentucky	(Wyandot)	"meadow land"
Manhattan	(tribal name)	"rocky island mountain"
Massachusetts	(Algonquian)	"great hills"
Menominee	(tribal name)	"wild rice people"
Mexico	(Aztec)	"place of the war god"
Miami	(Ojibwa)	"people of the peninsula"
Michigan	(Ojibwa)	"big lake"
Milwaukee	(Algonquian)	"good land"
Minnesota	(Siouan)	"water reflecting the sky"
Mississippi	(Algonquian)	"big river, father of waters"
Missouri	(tribal name)	"muddy waters"
Nantucket	(Algonquian)	"narrow tidal river"
Nebraska	(Siouan)	"flat, calm water"
Niagara	(Iroquois)	"thundering waters"
Ohio	(Iroquois)	"beautiful"
Oklahoma	(Choctaw)	"red people"
Omaha	(tribal name)	"upstream people"
Ontario	(Iroquois)	"sparkling waters"
Oregon	(Shoshone)	"place & river of plenty"

Ottawa	(Algonquian)	"place to trade"
Potomac	(Delaware)	"place to meet & trade"
Tennessee	(Cherokee)	"place of villages"
Texas	(tribal name)	"allies"
Utah	(tribal name, Ute)	"in the mountains"
Wisconsin	(Ojibwa)	"gathering waters & meadows"
Wyoming	(Delaware)	"great meadows"

Threads of ancient mound builders languages flow through some of these names. Perhaps most of these names come directly from ancient linguistic mound builders' societies. Many of these are Anglicized versions of the original Indian names or they bear Spanish or French influences in the current spelling. Some names bear specific tribal origins while others come out of a native language family, like the Algonquian or the Siouan, and are difficult to trace further.

We have a rich linguistic American Indian history and speak in 'native tongues' when we mention countless places in the western hemisphere. Many tribal names originally said "the people" or "the people of a particular place or resource." Indian place names are poetic reminders of an ancient, inspirational presence across the land, where millions of American Indian descendants continue to live.

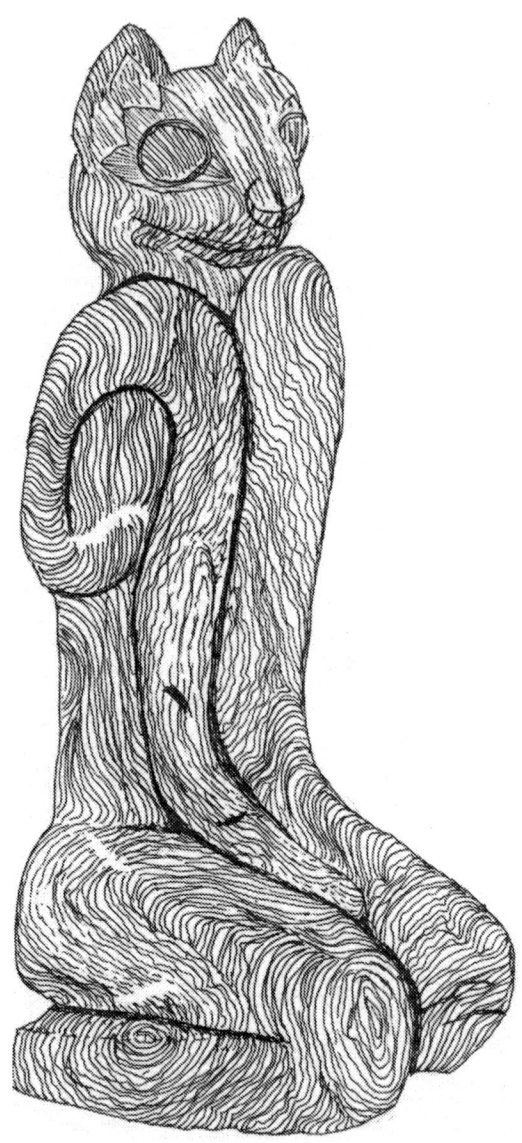

Southern Cult Florida Panther (Puma)--Carved by Calusa Indians of cedar wood; six-inches high. Found in 1884 on Key Marco Island off Florida's west coast; excavated in a pre-contact site.

Some Final Thoughts...

Centuries after the last mound builders, Meriwether Lewis wrote rhapsodically in the spring of 1805 of "seens of visionary enchantment," of coming upon "immence herds of Buffaloe, Elk, deer & Antelopes feeding in one common and boundless pasture." Lewis and Clark were exploring the regions along the Missouri River heading west. Their patron Thomas Jefferson thought it might take a thousand years to populate the newly acquired Louisiana Purchase. It required less than one hundred. Celebrating the 200th anniversary of the Purchase, Timothy Egan wrote in the New York Times, only "tattered scraps of once-splendid mosaics" remain.

> Look deep enough into the history of almost any Iowa town and you come to the primordial 19th-century tale of breaking the prairie as if it were a herd of wild horses.... The Iowa prairie was well and truly broken. The conventional figure is that 80 percent of the state was once prairie and that 99.9 percent of it is now gone, replaced by what used to be mixed farms and are now almost exclusively corn and soybean fields.
>
> —Verlyn Klinkenborg, N Y Times Editorial, 8/6/03

Here in the 21st century we reckon with the failure of some forms of conventional farming and acknowledge the momentous comeback of the American bison to the Great Plains as well as to farming initiatives beyond the prairies. It is a pleasant irony that American Indian farmers have been asked to take charge of hundreds of thousands of acres of high plains buffalo grazing lands because of their natural instincts and success. Some of these businesses will provide the buffalo meat to feed a hungry nation with a red meat deemed healthier than most others for certain heart patients. American farmers are also raising thousands of acres of American ginseng, prairie coneflower, evening primrose, sunflowers, sunchokes, jojoba, chaparral, opuntia, and various wild orchids to supply the growing health and healing fields. Wild harvested cranberries, blueberries, raspberries, hazelnuts, pecans, wildrice, and witch hazel continue to be lucrative businesses to meet growing demands for more natural foods and medicines.

To be successful futurists, we must look back with clarity over our history. We have a particularly rich and astonishingly diverse ancient history. America's relatively young civilization continues to face daunting problems and still has much to learn from her ancient forebears. America's "cradles of civilization" were many, dotted by compelling native societies living in close association with Mother Earth.

They were not perfect. Yet we search through the remnants for more clues to their greatness.

When The Moon is a Silver Canoe
Ballad © by E. Barrie Kavasch [*Remembering the Mound Builders*]
Sing in a lilting, happy 6/8 time

When the Moon is a Silver Canoe, sailing across the sky,
Peer up into the Sky World blue, watch the stars fly by.
Flickering campfires changing hues, Eagle circles on high.
Children gather to hear the news, as old stories begin to sigh;
    [refrain: *old stories begin to sigh, old stories begin to sigh,*
      when the moon is a silver canoe, old stories begin to sigh.

When the Moon is a Silver Canoe, gliding low across the sky,
Stories spill out into the blue, their characters start to cry,
Coyote howls a tale that's true, crow and raven aren't shy;
Shamans drum the dancers cue, while Spirits pray to the light;
    [refrain: *while Spirits pray to the light, Spirits pray to the light,*
      when the moon is a silver canoe, Spirits pray to the light.

When the Moon is a Silver Canoe, racing west across the night,
Pumpkins dance with beans and corn, trailing vines off out of sight.
Bats hunt darkly in squeaking flights, nightingales sing the blues;
Three Sisters share a warm embrace, their stories hold magical clues;
    [refrain: *their stories hold magical clues, stories hold magical clues,*
      when the moon is a silver canoe, stories hold magical clues.

When the Moon is a Silver Canoe, sailing boldly down the sky,
Star People chant the traveling songs and Spirits start to fly,
Stories shine as night winds sigh; come sail with me on high;
Ballads sing of earlier times when EarthMaker lived close by;
    [refrain: *when EarthMaker lived close by, EarthMaker lived close by,*
      when the moon is a silver canoe, EarthMaker lives close by...

Conclusions~How Did These Magnificent Cultures End? 213

> Grandmother Moon is Mother Earth's natural satellite illuminated by Grandfather Sun's reflected light, which varies during its 29 ½ day rotation around Mother Earth. Grandmother Moon is almost one eighth the size of Mother Earth, and she shows us a beginning and ending crescent phase each month when she is only partially illuminated by Grandfather Sun. These slim crescents hang like silver smiles or tipped silver canoes~seeming to beckon to the storytellers.

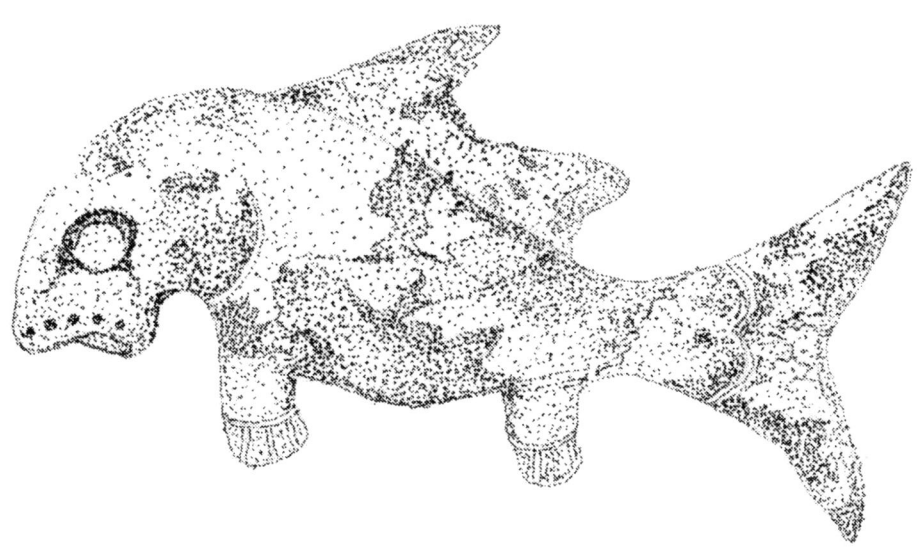

Hopewell Hammered Copper Redhorse Sucker or Buffalofish, found in Ohio; symbol of the Underworld.

Glossary

Adena Culture (1000-100 B.C.) The early Mound Builders, who flourished across the American midlands to the Atlantic coast.

Algonquian A family of languages spoken by many tribes in the Northeast and on the Plains and around the Great Lakes regions.

Ancestors Members of the same family or culture group from earlier generations.

Anthropologists Scientists who study earlier cultures.

Archaeologists Scientists who study the history of the art and cultural remains of past human life, usually by digging in specific sites.

Atlatl (from the Aztec & early Nahuatl) Device for spear-throwing.

Buffalo Bison are North American buffalo.

Burl A large growth on a tree, like a tumor. The growth pattern is swirled and more attractive within the burl. The outer bark is healing.

Cahokia Named for a local tribe of Illinois Indians, Cahokia is the largest Temple Mounds Complex in North America.

Ceremonial knives Usually large, ornamental blades pecked from fine stone, like flint or jade, obsidian, or alabaster.

Civilization The stage in the development of people when they live together in an ordered community or group of communities.

Clan A group composed of related families from several households, and an important unit of social organization among many North American Indians.

Cult A system or community of people sharing certain common interests and rituals.

Culture The ideas, art, beliefs, tools, and other characteristics of a particular civilization or social group in a certain time.

Earthen Mounds or earthworks Huge banks or mounds of earth that have been moved to form unique shapes.

Effigies Things that bear recognizable shapes and forms, such as a mound of earth shaped like a lizard, snake, or bear.

Egalitarian society One in which everyone is more or less equal in politics and law.

Excavate To dig up an ancient site, or something, to uncover more information about it.

Gorgets Pieces of armor protecting the throat; like ornamental collars.

Herbalists People who are skilled in the knowledge and use of herbs for healing and other special uses.

Hopewell Culture (200 B.C.-A.D. 500) A broad network of Mound Builders stretching from Mississippi to Minnesota, and Nebraska to Virginia.

Mississippians (A.D. 750-1500+) Hundreds of Mound Builders societies flourished throughout the Tennessee, Cumberland, and Mississippi river valleys.

Mound Builders (1000 B.C.-A.D. 1500) Advanced cultures of prehistoric Indians with sizable populations and sophisticated technologies who built thousands of mounds.

Prehistoric The time before written historical records; in North America it often refers to the time before Europeans arrived.

Sachems An Algonquian Indian word for their leaders, the men and women who ruled.

Shamans A Tungus (Siberian) tribal word for people who can journey into nonordinary reality to divine the cause of illness, search for game animals, and help spirits move on.

Technology The body of knowledge used by a culture to fashion its life and needs.

Temple Mound Builders (A.D.750-1500) the late Woodland Mississippians who built flat-topped mounds upon which their temples stood.

Tribe A community that shares cultural and social traits, and has a specific geographical location.

Woodland Indians Cultures who evolved special traditions more than 3,000 years ago that set them apart from the Archaic and Paleoindians before them.

Zoomorphic Possessing a distinct animal shape, such as a mound of earth or clay pot.

Illustrations

Ink Illustrations:

1. Great Serpent Mound in Ohio
2. Rock Eagle Stone Effigy Mound in Georgia
3. Butterfly weed
4. Bloodroot
5. Blueberry
6. Strawberry
7. Bee Balm
8. Cardinal flower
9. Evening Primrose
10. Bearberry
11. Wild Tobacco & Coyote Tobacco
12. Pipsissewa
13. Bearskin-clad Hopewell Shaman figure
14. Hopewell Beaver Effigy Pipe
15. Jasper Owl—Poverty Point, ca. 1500 B.C.
16. 3 Spider Design Shell Gorgets
17. Yarrow

18. Chippewa Clan Petition (Wisconsin)

19. Adena Male Shaman Pipe

20. Sweet Flag & Blue Flag

21. Sumac

22. Spiro Mound Copper Warrior

23. American Ginseng

24. Wild Onions & Garlic

25. Southern Cult objects

26. Hopewell Mica Hand

27. Hopewell Bird Talon Mica Foot

28. Hopewell Clay Woman & Child

29. Ceremonial Blade & stone hoe

30. Ohio River Valley Map

31. Mound Builders Regional Map

32. Southern Cult Panther carving

33. Effigy mounds objects

34. Mississippian Deer Mask

35. Hopewell Copper Hawk & Fish

36. Atlatl, Tennessee Banner Stones

37. Raptor & Man Stone Pipe

Additional illustrations

1. 2 Poverty Point site drawings by Jon L. Gibson

2. Cahokia Birger Figurine photograph by Illinois State Museum

3. Adena Mound detail, Grave Creek Mounds, Moundsville, WV

4. 3 Squier & Davis mounds sites drawings, 1848.

5. 4 mound builders cultural site maps, incl. De Soto's explorations. R. Silverberg.

For Further Information

Further reading:

AMERICA A.D. 1000: THE LAND AND THE LEGENDS, by Ron Fisher, National Geographic Society, Washington, D.C., 1999.

AMERICAN INDIAN SCIENCE: A NEW LOOK AT OLD CULTURES, by Fern G. Brown, Twenty-First Century Books, NY, 1997.

ASTRONOMY IN ANCIENT TIMES, by Isaac Asimov; "Library of the Universe" series, Gareth Stevens, Inc., NY, 1995.

BATTLEFIELDS AND BURIAL GROUNDS: THE INDIAN STRUGGLE TO PROTECT ANCESTRAL GRAVES IN THE UNITED STATES, by Roger C. Echo-Hawk and Walter R. Echo-Hawk, Lerner Publications, Minneapolis, MN, 1994.

BUFFALO BIRD WOMAN'S GARDEN, AS TOLD TO GILBERT L. WILSON: AGRICULTURE OF THE HIDATSA INDIANS. St. Paul; Minnesota Historical Society Press, 1987, 1917.

EXPLORING ANCIENT NATIVE AMERICA: AN ARCHAEOLOGICAL GUIDE, by David Hurst Thomas, Routledge, NY, 1999.

EXPLORING NATIVE NORTH AMERICA, by David Hurst Thomas; Oxford University Press, NY, 2000.

GREAT MYSTERIES: ANCIENT MYSTERIES, by Rupert Matthews; The Bookwright Press, NY, 1989.

HIDDEN CITIES: THE DISCOVERY AND LOSS OF ANCIENT NORTH AMERICAN CIVILIZATION, by Roger G. Kennedy; The Free Press, Macmillan, NY, 1994.

HOW DO WE KNOW WHERE PEOPLE CAME FROM?, by Mike Corbishley, "How Do We Know?" series, Raintree Steck-Vaughn, Austin, TX, 1995.

INDIAN MOUNDS OF WISCONSIN, by Robert A. Birmingham & Leslie E. Eisenberg; Madison: University of Wisconsin Press. 2000.

MEN AND MOUNDS IN ANCIENT AMERICA, by John Strong, NY; manuscript; 1980.

MOTHER EARTH, FATHER SKY: NATIVE AMERICAN MYTH, by Duncan Baird Publishers/Time-Life Books, NY, 1997

MOUNDS OF EARTH AND SHELL: NATIVE PEOPLES SERIES, by Bonnie Shemie; Simon and Schuster Children's, NY, 1994.

POVERTY POINT: A TERMINAL ARCHAIC CULTURE OF THE LOWER MISSISSIPPI, by Jon L. Gibson, Louisiana Department Of Culture, Recreation, & Tourism, 1996.

PYRAMID, by James Putnam; Knopf Books, NY, 1994.

SMOKING PIPES OF THE NORTH AMERICAN INDIANS. by J.C.H. King, London: British Museum Publications Ltd. 1977.

THE ANCIENT MOUNDS OF POVERTY POINT: Place of Rings, by Jon L. Gibson, Gainesville: University Press of Florida. 2000.

THE FIRST AMERICANS. by the editors of Time-Life Books; Alexandria, VA. 1992.

THE SACRED GEOGRAPHY OF THE AMERICAN MOUND BUILDERS, by Maureen Korp, Native American Studies Volume 2; Lewiston, NY:The Edwin Mellen Press, 1990.

UNSOLVED MYSTERIES: NATIVE AMERICAN MONUMENTS, by Brian Innes; Raintree Steck-Vaughn, Austin, TX, 1999.

WHEN THE MOON IS A SILVER CANOE: Legends of the Wisconsin Dells. By Capt. Don Saunders; Wisconsin: Baraboo Publishing Co., 1947.

1491:America Before Columbus was more Sophisticated and more Populous than we ever Thought-and a more Livable Place than Europe, by Charles C. Mann. *The Atlantic Monthly,* Vol. 289(3)41-53; 3/2002.

Videos:

MYTHS AND MOUNDBUILDERS: Explore the mysterious mounds left behind by America's Native People, Odyssey: PBS 263, Color, 58 minutes. Explores a fascinating range of theories and visits some of the most famous sites. Talks with the scientists excavating some of the major sites.

There are a growing number of good videos available…

Internet Web Sites to explore:

http://www.state.il.us/HPA/sites/CAHOKIAmounds01.HTM
http://www.maxwell.synedu/nativeweb/
http://www.si.edu/nmai
http://www.ohiohistory.org/places/moundbld/index.html
http://www.ohiohistory.org/places/serpent/index.html
http://www.ohiokids.org/ohc/archaeol/p_indian/tradit/~hswoodland.html
http://bio02.uthscsa.edu/aisesnet.html

Many additional sites, maps, and details are also available.

Mound Sites & Places to visit

Mounds and mounds complexes are located all across North America. Various Native American groups created sacred space at different times and in unique places. Much of this rests on private property now. The remnants we see today are special legacies from an ancient past that continue to imbue the earth with honor. They are stretched across 18 states. Many of these sites do not have formal addresses. Some sites are controlled within the National Park Systems, yet many more are on private property.

Alabama (1-800-ALABAMA) is the "Heart of Dixie" and home of the Yellowhammer woodpecker and the southern pines. It is also the homeland of the Creek, Cherokee, Choctaw, and Chickasaw Indians, and the following mounds sites:

Woodland Ceremonial Mound & Copena Burial Mound
1219 Co. Road 187
1-256-905-2494
Danville, AL
Woodland Indians ceremonial mounds covering 1.8 acres; the tallest standing 27 feet tall.

Oakville Indian Park & Museum
1219 Co. Road 187
1-256-905-2494
Danville, AL
Largest Woodland Indian ceremonial mound in region.

Indian Mound & Museum
1028 South Court Street
1-256-760-6427
Florence, AL
Large Mississippian mound 43 feet high.

Moundville Archaeological Park
1 Mound Parkway
1-205-371-2234
Moundville, AL
26 large prehistoric platform mounds in this 320-acre park include temple mounds near Black Warrior river.

Shell Mound Park
2 North Iberville Street
1-334-861-2882
Dauphin Island, AL
Prehistoric Indian mounds.

Arizona (1-888-520-3434) is the "Grand Canyon State" and home of the Cactus Wren and Saguaro Cactus; it is also the homeland of the Apache, Navajo, Tohono O'Odham (Papago), Pima, Mohave, Hualapai, Chemehuevi, and Hopi Indians, and the following:

Pueblo Grande Museum
Phoenix, AZ
1-602-495-4619
Large Hohokam mound site plus two prehistoric ball courts.

Arkansas (1-800-628-8725) is the "Land of Opportunity" and home of the Mockingbird and the Pine tree; it is also the homeland of unique prehistoric moundbuilders cultures, ancestors of historic Indian cultures of the rivers and plains, and the following:

Toltec Mounds State Park
US 165
1-501-961-9442
Scott, AR
Prehistoric earthworks built by 400 A.D.; complex site in the lower Mississippi River valley.

Florida (1-888-202-4581) is the "Sunshine State" and home of the Mockingbird and Sabal Palmetto Palm; it is also the homeland of the Seminole and Miccosuki Indians and their earlier ancestors, along with the following sites:

Crystal River State Archaeological Site
North Museum Point, off US 19
1-904-795-3817
Crystal River, FL
Diverse mound site just inland from the Gulf.

Indian Temple Mound Museum
FL 139
1-904-243-6512
Ft. Walton Beach, FL
Site of temple mound illustrating 10,000 years of Indian occupation on the northwest Florida coast.

Lake Jackson Mounds State Site
1-904-562-0042
2 miles north of I-10, at southern tip of Lake Jackson
Tallahassee, FL
Major mound and village site with 7 mounds and a plaza.

Georgia (1-800-VISIT-GA) is "Empire State of the South" and home of the Brown Thrasher and Cherokee Rose, and home of early Algonquian People, and these sites:

Etowah Indian Mound State Historic Park
SR 61/113, 6 miles south of I-75
1-404-387-3737
Cartersville, GA
Major Mississippian ceremonial center.

Kolomoki Mounds Historic Park
off US 27, 6 miles north of Blakely
1-912-723-5296
Blakely, GA
Large mound complex on 1,293 acres; 7 mounds, plus a museum charting Indian cultures in this region from 5,000 B.C.

Ocmulgee National Monument
1-912-752-8257
US 80E, from I-16 exit 4
Macon, GA

Spectacular mounds ceremonial center showing 10,000 years of Native American cultures, plus a museum.

Rock Eagle Stone Effigy Mound
no phone
Eatonton, GA
Enormous stone effigy mound of an eagle dating back 5,000 years.

Illinois (1-800-226-6632) is our "Prairie State" and home of the Cardinal and Native Violet, as well as homeland of the Illini and Kickapoo Indians and their earliest moundbuilding ancestors, and the following sites:

Cahokia Mounds State Historic Site
1-618-346-5160
30 Ramey Street, 6 miles east off I-55/70
cahokiamounds@ezl.com
Collinsville, IL 62234
Site of the largest city in prehistoric North America, established about 700 A.D. Late Mississippian complex. Monks Mound and Woodhenge celestial observatory.
www.cahokiamounds.com

Dickson Mounds Museum
1-309-547-3721
SR 97, 60 miles northwest of Springfield
Springfield, IL
Site of a large Mississippian community, 1100–1350 A.D.

Indiana (1-800-800-9939) is our "Hoosier State" and home of the Peony and Tulip Poplar, as well as homeland of the Wyandotte Indians and their ancestors, and the following sites:

Angel Mounds State Historic Site
1-812-853-3956
8215 Pollock Avenue
Evansville, IN
Best preserved Mississippian site, including temple and houses, dating from about 900 A.D.

Wyandotte Cave
1-812-738-2782
SR 62
Leavenworth, IN
Prehistoric quarry and sacred site. Source of Wyandotte chert.

Iowa (1-800-345-4692) is the "Hawkeye State" and home of the Eastern Goldfinch and Wild Rose, as well as the homeland of the Mesquakie, Sac and Fox Indians and their ancestors, and the following sites:

Effigy Mounds National Monument
1-319-873-2356
SR 76, 3 miles north of Marquette
Marquette, IA
Spectacular site of 200 mounds, including the "Marching Bears" and 26 animal effigy mounds, covering over 2,500 years of Effigy Mound Culture work.

Toolesboro Indian Mounds National Historic Landmark
1-319-523-8381
Hgwy 99
5-acre site protecting Hopewell burial mounds, plus a fine educational center & exhibits.
Wapello, IA

Kentucky (1-800-225-8747) is the "Bluegrass State" and home of the Cardinal and the Tulip Poplar as well as the homeland of the Cherokee and the earlier Mississippian culture, plus these special sites:

Mammoth Cave National Park
1-502-758-2251
10 miles west of Cave City, KY
Huge cave with more than 300 miles of underground passages; one of the longest cave systems known.

Wickliffe Mounds Research Center
1-270-335-3681
94 Green Street, US 51
Ceremonial & trade center site;

Wickliffe, KY
Mississippian village occupied more than 1,000 years ago.

Louisiana (1-800-33-GUMBO) is "Cajun country," the "Pelican State" and home of the Brown Pelican, Magnolia, and homeland of the Choctaw, Chickasaw, Houma, Tunica-Biloxi, Caddo, and Natchez Indians and their ancestors. Some of the special sites are:

Marksville State Historic Site
1-888-253-8954
837 Martin Luther King Drive
Marksville, LA
Earthen mounds dating from 400 A.D.

Poverty Point State Area
1-318-926-5492 or 1-888-926-5492
6859 SR 577, about 6 miles out of Epps, LA
povertypoint@crt.state.la.us
A 400-acre site of Archaic Indian earthworks dating from 1700 B.C.

Minnesota (1-800-657-3700) is the "Gopher State," home of the Common Loon and the Lady's Slipper Orchid, and was settled over 10,000 years ago by ancestors of today's Ojibwa (Chippewa) and other Great Lakes Indians. Some of the many special sites are:

Grand Mound Interpretive Center
1-218-285-3332
SR 11, 17 miles west of International Falls, MN
Largest burial mound in the Upper Midwest, almost 2,000 years old.

Indian Mounds Park
1-612-296-6157
East of the business district, overlooking the Mississippi River
St. Paul, MN
Several remarkable ancient mounds.

Lake Itasca State Park
1-218-266-2100
HC05, Box 4

Lake Itasca, MN
Visit the mighty river's source, plus Indian exhibits.

Pipestone National Monument
1-507-825-5464
At the north edge, outside of town
Pipestone, MN
Sacred site mined by ancestors of today's Indians, and continuing, for the red *catlinite* (pipestone). This site covers over 300 acres.

Mississippi (1-800-WARMEST) is the "Magnolia State," home of the Mockingbird and Magnolia; it was settled over 10,000 years ago by ancestors of the Choctaw, Chickasaw, Creek, and Natchez Indians, who created of the following sites:

Emerald Mound
1-800-305-7417
Natchez Trace Parkway
Built around 1400 years ago by ancestors of the Natchez Indians; second largest ceremonial mound in the U.S., covering nearly 8 acres.

Grand Village of the Natchez Indians
400 Jefferson Davis Boulevard
1-601-446-6502
Natchez, MS
National Historic Landmark, ceremonial mound center from 1200 until 1730 A.D.

Nanih Waiya Historic Site
1-800-467-2757
Coy Community
Philadelphia, MS
Ancient sacred mounds: legendary birthplace of the Choctaw Indians.

Natchez Trace Parkway
1-800-305-7417
RR1, NT 1143
Tupelo, MS
Headquarters, museum, and visitors center. This scenic byway began over 8,000 years ago as an Indian and buffalo trail. It stretches from Natchez, MS to Nash-

ville, TN, with many historic markers to stop and appreciate along the way, along with its incredible natural beauty.

Winterville Mounds Museum State Park
1-601-378-5559
2415 Highway, 1 North Drive
Greenville, MS
One of the largest Indian mound groups in the valley.

Ohio (1-800-BUCKEYE) is the "Buckeye State," home of the Cardinal and the Buckeye, and homeland of the Shawnee, Mohican, and Delaware Indians, and their ancestors. Some of the many special sites are:

Adena State Memorial and Hopewell Culture National Historic Park
P O Box 353
1-800-413-4118
Chillicothe, OH
This is one of the most amazing prehistoric mounds sites.

Flint Ridge State Memorial Museum
1-614-787-2476
SR 688, 3 miles north of US 40
Brownsville, OH
Site of the famous Ohio Pipestone quarry.

Fort Ancient
1-513-932-4421
SR 350, 7 miles off Middleboro Road
Lebanon, OH
A 100-acre Hopewell settlement site atop the bluff overlooking the Little Miami River.

Miamisburg Mound State Memorial
SR 725
1-614-297-2300
Miamisburg, OH
Huge Adena mound 68-feet high.

Mound Cemetery
no phone
On Fifth Street
Marietta, OH
30-foot high aboriginal mound, once part of an extensive earthworks complex.

Hopewell Culture National Historic Park
1-614-774-1125
Chillicothe, OH
Remarkable 13-acre complex with 23 interpreted mounds.

Newark Earthworks
1-614-344-1920
SR 79
Newark, OH
Mound Builders State Memorial including very large earthworks and numerous smaller mounds.

Seip Mound
1-614-297-2301
US 50, 3 miles east of town
Bainbridge, OH
This is a great Hopewell mound with related exhibits.

Serpent Mound
1-513-587-2796
SR 73, 4 miles northwest of town
Locust Grove, OH
Spectacular, enigmatic earthworks site from prehistory.

Story Mound
Adena site (800BC-AD 100)—not accessible. 20-feet tall
1982 Velma Avenue
Chillicothe, OH
The Ohio Historical Society, northwest of the city c/o Site Operations Department

Oklahoma (1-800-654-8240) is "Indian Country" and the "Sooner State" and home of the Scissor-tail Flycatcher and Mistletoe. It is home to more than 60 different American Indian tribes, and some of the many ancestral sites are:

Spiro Mounds Archaeological Park
1-918-596-2700
SR 9
Spiro, OK
Complex mounds linked to the Southern cult in Mississippian times.

Tennessee (1-800-462-8366) is the "Volunteer State," home of the Mockingbird and Iris, and homeland of the Cherokee and Chickasaw Indians and their ancestors. Some of the many special sites are:

Chucalissa Indian Museum
1-901-785-3160
Mitchell Road, next to Fuller State Park
National Historic Landmark
Memphis, TN
Interesting mound site more than 1,000 years old.

Mississippi River Museum at Mud Island
1-901-576-7230
125 North Front Street
tells the early geological story of the river.
Memphis, TN
18 galleries & many exhibits; lower river development story.

Old Stone Fort State archaeological Area
1-615-723-5073
US 41
Manchester, TN
A 2,000-year-old ceremonial site with mounds and walls.

Pinson Mounds State Archaeological Area
1-901-988-5614
460 Ozier Road, off US 45
Pinson, TN
One of the largest Hopewell mound groups in North America.

Shiloh National Military Park
1-901-689-5275
10 miles southwest of town
Savannah, TN
More than 30 Mississippian mounds built between 1100 and 1300 A.D. in this area.

Texas (1-800-452-9292) is the "Lone Star State," home of the Mockingbird and Bluebonnet, and homeland of many southern tribes including the Seminole, Cherokee, Caddo, Comanche, Cheyenne, Kiowa, Arapaho, Alabama, and Coushatta Indians, who were driven off their ancestral lands. Some of the many special sites are:

Alibates Flint Quarries National Monument
1-806-857-3151
Off SR 136, 34 miles northeast of town
Amarillo, TX
Significant Paleoindian quarry of this distinctive flint.

Caddoan Mounds State Historic Site
1-409-858-3218
SH 21, 6 miles southwest of town
Alto, TX
Two mounds and a 1000-year-old village.

West Virginia (1-800-225-5982) is the "Mountain State," home of the Cardinal and Sugar Maple, and homeland of eastern Algonquian tribes and their early ancestors. Some of the many special sites are:

Grave Creek Mound
1-304-843-4128 or 1-800-CALLWVA
801 Jefferson Avenue
grave.creek@wvculture.org
Moundsville, WV
This mammoth mound is a 70-foot-tall Adena structure.

Wisconsin (1-888-577-5052) is the "Badger State," home of the Robin and Wood Violet, and homeland of the Oneida, Ojibwa, Stockbridge-Munsee, Lac du Flambeau Chippewa, Winnebago, Hochunk, and other native people and their early ancestors. Some of the many special sites are:

Lizard Mound County Park
1-414-335-4445
On County road A West Bend, WI
Numerous amazing animal effigy mounds.

Sheboygan Mound Park
no phone
Off Panther Avenue
Sheboygan, WI
33 effigy mounds shaped like panthers and deer, constructed over 1000 year ago.

There are countless additional sites spread across North America that are on private property, and visitors are usually not welcomed without permission.

Author's Profile

Elizabeth Barrie Kavasch was born in the heart of Mound Builders territory in Ohio more than half a century ago. She is the mother of two and grandmother of four wonderful, colorful, bright beings. She is an author, illustrator, and poet, as well as shamanic practitioner. Along with many years of significant research on the mound builders, she continues to make periodic shamanic journeys back to the diverse mound builders and their sites to visit with their willing spirits, especially the distinctive shamans, and others eager to share their information.

She is of American and American Indian descent, and is a 15th generation direct descendant of Pocahontas through her father's lineage. This is her 25th book. She is perhaps best know for her early books, *NATIVE HARVESTS* (Institute for American Indian Studies, 1998), *A STUDENT'S GUIDE TO NATIVE AMERICAN GENEALOGY* (Oryx, 1996), *EARTHMAKER'S LODGE* (Cobblestone, 1994). Her more recent books are *ANCESTRAL THREADS* (Iuniverse, 2003), *THE MEDICINE WHEEL GARDEN* (Bantam 2002) and *AMERICAN INDIAN HEALING ARTS* (Bantam, 1999.)

Barrie received the G.B. Foster Award for **Lifetime Achievements in Art & Literature in 2000** from the Herb Society of America. *AMERICAN INDIAN HEALING ARTS* was a finalist for the "**Best Books of 2000**" in New York City. Her book EARTHMAKER'S LODGE received the **Gold Star Achievement Book of the Year** from the Institute for American Indian Studies in Washington, CT, and the **Boston Book Binders Award**, and was selected by the state of California as a curriculum guide in their schools. She is a gifted storyteller and songwriter.

Barrie is also an energy healer and certified Reiki Master. She continues to explore the benefits of "sound healing" along with the use of Tibetan and Crystal Singing Bowls, and Peruvian Whistling Pots, along with other unique instruments.

Barrie is currently developing a series of Y/A novels based within early mound builders societies. Each novel explores the mysteries of an ancient mound building group and the range of fantastic phenomena these people were capable of developing. From shamanic journeywork to realms of heightened awareness, prophetic dreams, magic, and mysticism—these ancient folks held seemingly unlimited powers.

Index

A
Adena culture 26, 32, 54, 61, 62, 215
animal mounds, *see* places & sites to visit
anthropologists xxviii, 215
archaeologists xxvi, xxviii, 1, 2, 29, 44, 60, 79, 80, 92, 122, 129, 215

B
ballads 212
   When the Moon is a Silver Canoe 9, 212, 224
bear mounds, *see* places & sites to visit
burial mounds, *see* places & sites to visit

C
Cahokia Mounds, *see* places & sites to visit
cultures vii, xxi, 1, 2, 10, 12, 14, 15, 19, 25, 26, 27, 31, 32, 34, 35, 36, 43, 44, 52, 69, 93, 99, 104, 115, 116, 118, 126, 156, 163, 175, 180, 189, 195, 201, 215, 216, 217, 223, 228, 229, 230
   Adena 26, 32, 43, 52, 53, 54, 55, 56, 57, 58, 59, 61, 62, 63, 67, 195, 215, 220, 221, 234, 235, 237
   Archaic 3, 4, 5, 6, 7, 10, 11, 13, 14, 15, 19, 26, 32, 52, 94, 106, 194, 217, 224, 232
   Effigy Mound Builders xv, 8, 29, 80, 85, 89
   Hopewell vii, 22, 26, 27, 29, 32, 38, 43, 62, 63, 65, 66, 67, 68, 69, 70, 71, 72, 73, 75, 76, 78, 79, 80, 95, 112, 184, 195, 213, 216, 219, 220, 231, 234, 235, 236
   Mississippian 26, 30, 33, 43, 47, 80, 86, 87, 90, 93, 94, 95, 97, 98, 100, 102, 116, 117, 118, 126, 127, 136, 140, 149, 175, 182, 183, 184, 197, 198, 220, 227, 229, 230, 231, 232, 236, 237
   Paleoindian 1, 2, 3, 4, 32, 237
   Southern Cult, Southern Death Cult 45, 46, 199, 200, 210
   Temple Mound Builders 30, 31, 35, 47, 78, 93, 95, 98, 100, 123, 126, 163, 195, 217
   Woodland Indian 19, 20, 23, 27, 32, 41, 93, 112, 116, 191, 206, 217, 227

E
Eagle Mound, *see* places & sites to visit
effigy mounds
   Bear Mounds 88, 89, 231
effigy mounds, *see* places & sites to visit
   Eagle Mound 86
   Lizard Mound 88, 160, 194, 238
   Serpent Mound 29, 90, 91, 92, 219, 235

F
fiber and cordage 60, 107
Fort Ancient, *see* places & sites to visit

G
games 113, 150, 172
Grave Creek Mounds, *see* places & sites to visit
Great Serpent Mound, *see* places & sites to visit

H
Historians xxiii, xxvi, xxviii, 26
Hopewell culture 26, 32, 38, 62, 67, 69, 78, 195, 216, 234, 235

I

Illinois vii, xxiv, 7, 33, 62, 65, 67, 80, 94, 95, 97, 99, 102, 129, 205, 208, 215, 221, 230

Indiana 12, 54, 118, 230

Indians vii, ix, xv, xxi, xxii, xxiii, xxiv, xxvi, 6, 7, 8, 10, 14, 19, 20, 21, 23, 25, 27, 34, 39, 46, 47, 49, 53, 55, 72, 77, 86, 90, 93, 95, 97, 98, 100, 103, 104, 106, 112, 114, 116, 122, 127, 154, 158, 161, 164, 165, 167, 179, 183, 184, 190, 191, 199, 200, 201, 202, 205, 210, 215, 216, 217, 223, 224, 227, 228, 230, 231, 232, 233, 234, 236, 237

Iowa 29, 79, 86, 88, 208, 211, 231

J

Jefferson, Thomas xxiv, 211

L

Lizard Mound complex, *see* places & sites to visit

M

Mammoth Mound, *see* places & sites to visit

medicine wheels 39

Mississippi river 7, 10, 11, 15, 20, 30, 34, 47, 78, 79, 93, 126, 165, 170, 176, 202, 206, 216, 228, 232, 236

Mississippians 30, 43, 46, 51, 78, 80, 86, 93, 94, 95, 98, 126, 182, 183, 189, 195, 198, 216, 217

Missouri river 95, 106, 116, 117, 205, 211

Mound Builders ii, vii, xv, xviii, xxi, xxii, xxiii, xxiv, xxvi, 7, 8, 10, 12, 20, 25, 26, 29, 30, 31, 32, 35, 36, 37, 38, 40, 41, 43, 44, 45, 47, 50, 51, 52, 55, 70, 78, 80, 81, 85, 86, 87, 89, 92, 93, 94, 95, 98, 100, 103, 104, 105, 106, 107, 108, 110, 111, 112, 115, 116, 119, 123, 126, 127, 128, 130, 136, 140, 144, 149, 150, 151, 153, 156, 158, 159, 161, 163, 164, 165, 166, 167, 168, 169, 170, 171, 172, 175, 177, 180, 182, 183, 187, 189, 190, 191, 193, 195, 199, 200, 201, 207, 209, 211, 212, 215, 216, 217, 220, 221, 224, 235, 239

Adena 26, 32, 43, 52, 53, 54, 55, 56, 57, 58, 59, 61, 62, 63, 67, 195, 215, 220, 221, 234, 235, 237

Effigy vii, xv, xxvi, xxvii, 8, 11, 29, 32, 43, 45, 53, 69, 72, 79, 80, 81, 83, 85, 86, 87, 88, 89, 90, 96, 102, 103, 105, 106, 183, 184, 185, 189, 194, 195, 206, 219, 220, 230, 231, 238

Hopewell vii, 22, 26, 27, 29, 32, 38, 43, 62, 63, 65, 66, 67, 68, 69, 70, 71, 72, 73, 75, 76, 78, 79, 80, 95, 112, 184, 195, 213, 216, 219, 220, 231, 234, 235, 236

Southern Cult 33, 34, 37, 45, 46, 48, 100, 152, 159, 199, 200, 210, 220, 236

Mound City, *see* places & sites to visit

Mounds vii, xv, xix, xxi, xxii, xxiii, xxiv, xxvi, xxvii, xxix, 10, 11, 12, 14, 19, 21, 25, 26, 27, 29, 30, 32, 33, 34, 36, 38, 39, 43, 44, 45, 46, 48, 52, 54, 55, 56, 57, 59, 61, 62, 63, 64, 67, 68, 69, 70, 71, 72, 75, 78, 79, 80, 81, 83, 85, 86, 87, 88, 89, 90, 93, 95, 97, 98, 99, 100, 102, 103, 104, 105, 106, 107, 111, 112, 114, 115, 120, 123, 126, 137, 140, 147, 148, 149, 150, 152, 160, 163, 165, 175, 176, 189, 194, 198, 199, 200, 201, 206, 215, 216, 217, 220, 221, 224, 225, 227, 228, 229, 230, 231, 232, 233, 234, 235, 236, 237, 238

Bear xvii, 4, 5, 8, 31, 41, 72, 76, 88, 90, 125, 130, 141, 142, 143, 146, 156, 162, 164, 171, 202, 209, 216

Burial vii, xxiv, xxvi, xxvii, xxviii, 19, 26, 29, 38, 39, 44, 48, 51, 53, 55, 56, 59, 60, 61, 64, 67, 69, 74, 75, 89, 98, 99, 102, 104, 105, 106, 107, 123, 127, 142, 149, 152, 175, 177, 194, 223, 227, 231, 232

Eagle xvii, 35, 55, 82, 83, 86, 87, 88, 89, 96, 100, 102, 212, 219, 230

Effigy vii, xv, xxvi, xxvii, 8, 11, 29, 32, 43, 45, 53, 69, 72, 79, 80, 81, 83, 85, 86, 87, 88, 89, 90, 96, 102, 103, 105,

106, 183, 184, 185, 189, 194, 195, 206, 219, 220, 230, 231, 238
Grave Creek 26, 56, 57, 59, 221, 237
Great Serpent 29, 90, 91, 219
Lizard 88, 160, 194, 216, 238
Mammoth 1, 2, 34, 59, 115, 200, 231, 237
Monk's Mound, *see* places & sites to visit
Moundville, *see* places & sites to visit
Poverty Point vii, 10, 11, 12, 13, 14, 26, 32, 52, 106, 111, 219, 221, 224, 232
Spiro 33, 45, 46, 51, 100, 102, 200, 220, 236

museums, *see* places & sites to visit

O

Ohio vii, xv, xxi, xxii, xxiv, 7, 12, 26, 28, 29, 33, 51, 52, 53, 54, 55, 57, 59, 62, 67, 68, 69, 70, 71, 73, 75, 76, 90, 91, 102, 111, 112, 118, 176, 205, 208, 213, 219, 220, 234, 235, 239

P

platform mounds xxii, xxiv, 45, 46, 228
poetry ii, 200
    Chickasaw Song 92
    Did They Hold The Vision? 194
    Portage la Prairie 205
    When the Moon is a Silver Canoe 9, 212, 224
    William Cullen Bryant xxiii
Poverty Point Mounds 32

R

reflections 9, 35, 130, 168, 177, 182, 190, 191, 201
    A Cahokia Childhood 151
    A Fiber Maker 130
    A Healer's Dream 160
    A Midwife's Tale 172
    A Shaman's Dream 161
    Adena, Graves Creek 52, 59, 60, 237

Cahokia 30, 33, 35, 51, 80, 94, 95, 98, 99, 100, 112, 116, 117, 118, 119, 120, 122, 123, 124, 125, 126, 127, 128, 129, 130, 136, 137, 140, 141, 143, 146, 147, 148, 149, 150, 151, 152, 162, 163, 165, 169, 175, 177, 178, 179, 180, 188, 189, 192, 195, 196, 197, 198, 199, 215, 221, 230
Cahokia, Scioto 230
Effigy, Eagle Town 29, 82, 83
Hopewell, Mound City 9, 62, 63, 65, 66, 67
My First Vision Quest 170
Temple, Cahokia 93, 94, 95, 98, 99, 100, 101, 116, 117, 139, 198, 199, 230
The Hummingbird Society 177
They Danced me out of Childhood… 168, 169

rivers xxi, xxii, 3, 4, 7, 20, 25, 31, 53, 54, 68, 70, 71, 89, 93, 95, 106, 115, 116, 117, 118, 164, 190, 194, 197, 205, 228
Rock Eagle Effigy Mound 87, 230
Russell Cave 14, 26, 32, 52

S

sacrifices 49, 122, 163
shamanism 154, 159, 162
shamans vii, 45, 55, 83, 85, 88, 119, 124, 128, 140, 141, 154, 156, 160, 161, 187, 196, 212, 216, 239
songs 129, 146, 149, 151, 154, 171, 175, 212
    When the Moon is a Silver Canoe 9, 212, 224
Spiro Mounds 33, 46, 236
stones and rocks 112
stories 4, 8, 9, 21, 42, 45, 46, 60, 66, 75, 85, 112, 147, 151, 169, 189, 190, 191, 207, 212
    A Story of the Pleiades 206
    The Choctaw Sacred Home 77
    The First Fire 41, 42
    The GreenWater Spirit 8, 9

The Lilies of the Lake 190
The Stone People 112, 113

T

technology 43, 103, 107, 184, 216
Temple Mound Builders 30, 31, 35, 47, 78, 93, 95, 98, 100, 123, 126, 163, 195, 217
Temple Mound Builders, *see* places & sites to visit
Temple Mounds vii, xxvi, 33, 48, 81, 102, 105, 120, 149, 165, 215, 228

Timeline 31, 32
totems 29, 89, 148, 203

W

Wisconsin xv, xxi, xxiv, xxvi, xxvii, 8, 9, 29, 79, 81, 83, 85, 86, 88, 89, 93, 126, 194, 199, 202, 203, 204, 205, 206, 209, 220, 224, 238
Woodland cultures 19, 44
    Eastern Woodland Indians 93, 191

0-595-30561-X

CPSIA information can be obtained at www.ICGtesting.com
Printed in the USA
LVOW061137290412

279585LV00004B/3/A